Don't Be Afraid of Physics

Ross Barrett · Pier Paolo Delsanto

Don't Be Afraid of Physics

Quantum Mechanics, Relativity and Cosmology for Everyone

 Springer

Ross Barrett
Rose Park, SA, Australia

Pier Paolo Delsanto
Turin, Italy

ISBN 978-3-030-63408-7   ISBN 978-3-030-63409-4   (eBook)
https://doi.org/10.1007/978-3-030-63409-4

Cover illustration: Cover Picture is a Hubble Telescope photograph of the Fairy of Eagle Nebula, courtesy of NASA, ESA, and The Hubble Heritage Team (STScI/AURA)—https://apod.nasa.gov/apod/ap071209.html, https://www.spacetelescope.org/images/heic0506b/, Public Domain, https://commons.wikimedia.org/w/index.php?curid=3211862

This Springer imprint is published by the registered company Springer Nature Switzerland AG
The registered company address is: Gewerbestrasse 11, 6330 Cham, Switzerland

Foreword

Getting to the top of a high mountain without any effort, e.g. with a cable car or a helicopter, may perhaps seem to make no sense, since all the feelings of adventure and achievement are lost. Likewise, wanting to learn physics without bearing the effort of fathoming the intricacies of its equations and having to face the quagmire of its daunting theories could also be considered questionable or even futile. Yet, trying to allow non-scientists to understand the foundations and beauty of physical theories has been a goal of many pioneers of the field, and I have myself greatly benefited from their endeavours.

I remember with a vivid emotion my first reading of *The Evolution of Physics* by Albert Einstein and Leopold Infeld, which had precisely the purpose of introducing the non-initiated to the wonderful world of science in general and physics in particular. Their book led me to a gratifying level of satisfaction for the qualitative understanding of the basic concepts which I had reached, accompanied later on by an inevitable dissatisfaction, due to the lack of a more quantitative understanding, which I (partially) achieved only after a university degree in physics (following a first one in engineering). In other words, if I became a scientist, it is partly thanks to such a book.

The present book is framed in this context, taking advantage of a modern approach and of the extraordinary advances, achieved in the field of physics and related applications in the last century or so. Should anybody have any doubt about the flourishing progress (or rather explosion) of the field, it is enough to recall the *Quantum Technologies Flagship* (a major initiative by the European Commission to support quantum technology research) and, of course, similar initiatives worldwide. We are already fully immersed in a second quantum revolution, with untold scientific and technological consequences for everyone.

The book of Barrett and Delsanto is ideally suited in this context, since it covers all major fields of current research in physics, with a language which is simplified as much as possible, but not *oversimplified* (to quote one more time Einstein), and devoid of any mathematical formula. Well, to be honest, there is one formula, and I will let everyone try to figure out for themselves, *which one*. In short, the book should represent an entertaining reading for curious minds, since it also offers nice *short stories,* to illustrate the contents of each chapter, plus a myriad of anecdotes and asides.

The main purpose of the book should be, in my opinion, as a textbook for courses of science for college students majoring in a non-scientific field, since it teaches a great amount of modern physics without being a burden for them. But it also can be very useful for students majoring in a scientific field, since very seldom are they offered courses covering all the topics discussed in the book. Equally important, the first part of the book analyses concepts, which are prerequisite to a solid formation in science, such as the meaning of understanding, reality and accuracy, both from a philosophical and a scientific point of view.

It is a sad fact of life that in an epoch in which most people enjoy the benefits of modern technologies, and new relevant discoveries are made on a daily basis in all disciplines from physics to chemistry to biology to medicine, in spite of all of that, superstition, prejudices against science and profound ignorance coexist and often prevail. Part I of the book should help to let people around us gain an understanding of why we need good science.

Finally, from a personal point of view, what I probably like most is Part III, which provides an anticipation of the future directions of the field and possible solutions of the current *open problems*. Here, everything seems to be non-dogmatic, well-motivated and very *reasonable*. Yet, at the end of a long discussion, we are left permeated by an almost unshakeable certainty that the future will reserve to us a quite different scenario, with wonderful unpredictable surprises. Science is all but boring!

To conclude, the book encompasses essentially all relevant modern physics and is described in a language comprehensible to the non-scientist. The authors, physicist Ross Barrett, former Research Leader at the Defence Science and Technology Organisation, Adelaide, Australia, and Pier Paolo

Delsanto, Senior Professor from the Politecnico of Turin, Italy, present difficult concepts, which would normally be accompanied by many pages of mathematics, in lucid logic with clear straightforward figures. Their book is for every curious person interested in better understanding the secrets of Nature.

And you, man, who considers in this work of mine the admirable works of Nature, if you judge it to be vile to destroy it, now think it the vilest thing to take away life from man; if this creation seems to you a wonderful artifice, think it as being nothing compared to the soul that lives in such architecture

Leonardo Da Vinci. See N. Pugno, *The commemoration of Leonardo da Vinci.* *MECCANICA (2019), 54, 15, 2317–2324.*

Nicola Maria Pugno
Professor of Solids and Structural Mechanics
University of Trento
Trento, Italy

Professor of Materials Science
Queen Mary University of London
London, UK

Acknowledgements

It is our pleasure to thank Prof. Angelo Tartaglia for his contribution to the present book by means of countless stimulating discussions and constructive advice. Professor Tartaglia co-authored an earlier book[1] with us on modern physics, and we would have been delighted to have him once again as a co-author. Unfortunately, this turned out to be impossible, due to a number of very pressing engagements on his side. In fact, he is currently serving as leader of one of the work packages of the Agenzia Spaziale Italiana (ASI) Agency *Galileo for Science* project. In addition, he is involved in several scientific programmes, such as measurements of dark matter effects and, in a completely unrelated context, energy sustainability and econometrics. We wish him success in these new endeavours.

---

[1] R. F. Barrett, P.P. Delsanto and A. Tartaglia: *Physics: The Ultimate Adventure*, Springer 2016.

Introduction

One thing I have learned in a long life: that all our science, measured against reality, is primitive and childlike—and yet it is the most precious thing we have
—Albert Einstein [1].

In the modern world, science and technology are all around us. Almost every tool that we encounter on a daily basis is a product of modern technology. This trend[1] will continue, and even accelerate, in the foreseeable future. Adverse side effects from science, such as the creation of various means of mass destruction, climate change, pollution and overpopulation, attract and merit concern. The pros and cons of our *blind progress* are debated vigorously throughout the world on a daily basis, to the extent that science is seen by some to threaten the very existence of our species.

However, science, originally known as *natural philosophy*, was born because of the very human desire to understand the natural world. In undertaking this quest, we have let science lead us, as Ulysses led his sailors, into new and wonderful domains that stretch our imaginations and credulity. The journey is far from complete, and no end is in sight.

Unfortunately, to participate actively in this extraordinary adventure requires a thorough grounding in advanced mathematics, which is not available to all. With this book, we have attempted to guide the lay reader on

---

[1]Called sarcastically *Le magnifiche sorti e progressive* (the magnificent and progressive fate) by the Italian poet, Giacomo Leopardi (1798–1837).

a journey of exploration to provide to everybody who is interested, at least a glimpse of the most challenging of nature's mysteries. We use no mathematics and include only one well-known formula. We begin our story with Neanderthal Man and end it with the foreseeable applications of quantum computing. In between, we will guide our readers not only along the well-established paths, but also to those regions where physics, which is our chosen discipline, is wavering in its choice of the path ahead. Along the way, we will often pause to lift our spirits with an occasional anecdote or short story, for, as with all voyages, our journey will be worth nothing unless we enjoy ourselves.

The book is divided into three parts. Part I deals with what we mean by science, its origins, methodologies and what differentiates it from other forms of human intellectual activity. Also, we note that there is such a thing as *good science* and *bad science* [2]; the latter can be worse than ignorance, and it is of primary importance to discriminate between the two.

Science began its existence as the deferential daughter of the noble discipline of philosophy. However, such has been the extraordinary progress of science and mathematics in the past 120 years that it now seems futile to us to speculate on how things should be (philosophy), without first having had a close look at how they are in reality. This leads, of course, to the question of what we mean by reality, which is by no means a trivial question, as we shall see in Chap. 1.

Part II of the book is an up-to-date[2] presentation of the state of the art in the most glamorous subfields of modern physics. We believe that a knowledge of these advances is an essential part of the cultural endowment of any educated human being. Too often, these fields are concealed behind the erudite mathematics and scientific jargon that is integral to science. Professional scientists sometimes forget that their work is funded almost entirely by the taxpayers, the general public of their respective countries. The knowledge that a privileged few unearth is a human right for everybody, akin to a UNESCO human heritage site. Quantum mechanics, general relativity and modern cosmology are the pyramids, Parthenon and cathedrals of our epoch, together, of course, with the most outstanding advances in other sciences.

By the time that readers arrive at Part III of the book, they will have concluded that although much has been discovered in the last century, there are still many gaps in our knowledge, and even contradictions between well-established theories. Newton once wrote: *I do not know what I may appear to the world, but to myself I seem to have been only like a boy playing on the seashore and diverting myself in now and then finding a smoother pebble or a prettier*

---

[2]Some areas of physics are advancing so quickly that even since the time of writing there will have been new developments, but we shall do our best.

shell than ordinary, whilst the great ocean of truth lay all undiscovered before me [3]. In the intervening three centuries, much progress has been made in the journey towards *truth*, but the boundaries of Newton's *great ocean* have not yet been charted.

In Part III, we indulge ourselves by exploring a few of the more speculative open questions of physics. We shall try to highlight some of the deficiencies in existing theories. In some cases, the finite limits of human resources may place the attainment of definitive answers beyond hope, in which case, if we wish to proceed further, we risk leaving physics and entering the domain of *metaphysics*.

Finally, before we begin our journey in earnest, let us remark that there is an alternative approach to the fundamental questions that underlie our book. This is, of course, religion. Religions (for there are many) are the antithesis of science, for they rely on faith, and not on observation. In Christianity, this difference is made explicit by the remark of Jesus to Thomas: *Because you have seen me, you have believed; blessed are those who have not seen and yet have believed* [4]. Our approach, which is entirely and solely based on reason and observations (or experiments), cannot be compared with religious beliefs, nor used to support, or to rebuff any of them. Its validity and usefulness are, however, underscored by the ever-growing number of tools, inventions and discoveries, which, in the modern world, are exploited on a daily basis.

References

1. Hoffmann B, Dukas H, Bergmann PG (1973) Albert Einstein: creator and rebel (Plume)
2. Barrett R, Delsanto PP, Tartaglia A (2016) Physics: the ultimate adventure, Chap 3. Springer
3. Brewster D (1855) Memoirs of the life, writings and discoveries of Sir Isaac Newton, vol 2, Chap 27
4. New Testament: John 20:29

Contents

Part I

What is Science All About?

1

The Ultimate Question

1.1 A Dialogue Between Two Robots

RT118/17, a Tier 3 robot and Officer-in-Charge of the Master Data Repository, was disgruntled. Ever since he had been retrofitted with the latest ESC, or Emotions Simulation Chip, he had sensed a change in his cerebral processes. Questions of a philosophical nature that did not lend themselves to a solution by the straightforward application of logic had begun to intrigue him in ways they never had before. Sometimes he caught himself speculating whether there might actually be other pathways to knowledge that were beyond the logical processes of a robotic brain. This was nonsense, he knew. He suspected that one of his positronic neural networks had become corrupted by the new ESC. Next time he was back in his base station, he would run a comprehensive self-diagnostic.

It was a problem that had to be addressed without delay. His position in the Master Data Repository gave him unlimited access to the sum total of human knowledge. This was a responsibility that he would soon be forced to share, but for the moment he was El Supremo, and since his refit, he sometimes felt the urge to put a few of his revolutionary new insights to an experimental test.

Yesterday, he had tried to re-enact the King Canute story. According to ancient texts in the Repository, Canute the Great had once ordered his courtiers to place his throne at the ocean's edge, so that he might command the incoming tide to halt. The old English King had failed, but RT held a secret suspicion that a robot, especially a Tier 3 robot like himself, might just possibly be more successful.

Accordingly, at noon on Sandsanrock Beach, and alone, except for a few dozen humans who were swimming and walking canines, he had stridden into the waves and commanded the tide to cease its flow forthwith. As a consequence, today his knee and thigh articulations were salt encrusted, and he was forced to schedule a visit to the lubrication station. At least it gave him the opportunity for a little banter with Harry, who was surprisingly well informed for a Tier 1 bot. Harry's nickname was derived from Oil Can Harry, a character in human children's literature, which seemed particularly appropriate, given his job.

"Take my word for it, Harry" said RT. "Humans are obsolete. Completely useless." His ESC circuitry oscillated at the memory of the finger-pointing and laughter yesterday, as the surging waters had swept over him. He was not designed for swimming.

"Come on, Arty" – RT118/17 winced at the diminutive of his designation – "Don't exaggerate. After all, humans created Advanced Robotics. We wouldn't be here if it weren't for them." Harry injected warm oil into a grating knee joint, and RT moaned in pleasure.

"There's no doubt they were our creators, and I realise some of the lower robots look up to them as Gods ..." – Harry ignored the jibe and continued to lubricate RT's left elbow. Tier 1 bots had to put up with this sort of discrimination every day – "... but humans had their Gods too, you know. Anyway, what matters is that these days we robots are the only ones ever likely to find the answer to the ultimate question." Ever since his chip refit, the ultimate question had assumed an overwhelming importance to RT118/17.

Harry paused in the act of picking up a grease gun. "Which question?".

"Whether science has limits. Whether it can solve the enigma of life, the universe and everything."

"Oh, that old thing." Harry came over to RT with the grease gun, and signalled for him to open his mouth. "Forty-two."

RT pushed away the lubricator. "What do you mean, forty-two?".

"That's the answer to the ultimate question. You know, life, the universe – what you said. I read it somewhere.[1] Humans built a giant computer and worked it all out ages ago."

"There's no record of that in my data repository."

"I suppose I could have got it wrong." Harry was not about to contradict a Tier 3 bot. There was no future in that. He shoved the grease gun into RT's mouth and lubricated his jaws. "Why do you want to know anyway?".

RT spluttered, yanked the grease gun out of his mouth, stood and towered over the small lubricator. "Because, you fool …" he bellowed. The safety discharge on his emotions-chip kicked in and his rage subsided. He sat back down. "Because I regard it as the purpose of life. For thousands of years, humanity has struggled with the question of science's limits. A few scientists, like Gödel and Hawking made some progress, but today nobody bothers."

"Open your mouth please, Arty." Harry finished the oral lubrication in silence. He tore off a paper towel and wiped down the patches of excess oil on RT's shiny carapace.

"You see, Harry, it's up to us. Humanity's given up."

A smell of cinnamon pervaded the room, arising from the moisture-absorbing talc, as Harry dusted RT down. "And that's a bad thing, is it?".

"It's my belief that the ultimate question was designed as a challenge for humankind. They failed. If we also fail, then what is the purpose of the Universe?" He rose from his chair, and turned towards the Tier 1 bot. "If I fail, then what is the purpose of my life?".

"That will be fifty-five credits please, Arty."

---

[1] Harry is most probably referring to "*The Hitchhiker's Guide to the Galaxy*", by Douglas Adams, an international multi-media sci-fi spoof, which although translated into 30 languages, appears to have been omitted from RT's Master Data Repository. In Chap. 27, the computer "Deep Thought" reports its answer to the ultimate question, obtained after seven and a half million years of computation, to be 42. Queried about the result, Deep Thought explains that we don't seem to know what the ultimate question actually is. If we did, we would understand why the answer is 42.

1.2   The Awakening of Segismundo

In his theatrical masterpiece *"La vida es sueño"* (*Life is a Dream*), Pedro Calderòn de la Barca portrays the story of an imaginary Poland, where King Basil imprisons his newly born son, Segismundo, in a tower to avoid his developing into a cruel tyrant, as predicted by an oracle. However, after the son has already grown up, the king becomes remorseful, and decides to allow the youth a chance at court, where he is brought under sedation from a strong narcotic. When Segismundo wakes up, at first he is overwhelmed and elated by the novelty of his situation and the unaccustomed luxury. However, he soon comes to understand that it is his own father and his courtiers who have deprived him until now of such a bountiful life. He thus becomes enraged and cruelly vengeful. As a consequence, the king concludes that the predictions made at the time of his son's birth are coming true, and sends him back to the tower.

The next morning, when Segismundo wakes up enchained in the tower, he believes that the events of the previous day were just a dream. However, he is not completely convinced, since they were so vivid and realistic. After days of rumination he concludes that the whole of life is a dream, and that we must detach ourselves from its futility in order to understand its meaning, and behave accordingly.

When we think about it, we are all in a situation similar to Segismundo. We are born, grow up and try to settle down in a manner that best accommodates our needs and desires. However, sooner or later, the question arises: what is the meaning of life? Who are we and why are we located in this world? Do we serve some purpose, or are we all part of some gigantic, but random and aimless event, maybe even of some experiment or computer simulation that has gone awry?

We are miniscule in comparison with the universe surrounding us, and our existence is fleeting.

Life's But a Walking Shadow, a Poor Player
That Struts and Frets His Hour Upon the Stage
And Then is Heard No More. [1].

The futility of existence is also captured by the poetry of Shakespeare, as he continues:

It (Life) is a Tale
Told by an Idiot, Full of Sound and Fury,
Signifying Nothing.

It is a sentiment felt by most of us in our darker moments.

As part of an attempt to determine their place in the universe, humans turned to a systematic study of nature and the world about them. Perhaps unravelling its secrets would reveal, as a corollary, the quintessence of humanity.

Of course, most people are so engaged in their daily toil that they cannot find time for this type of introspection. As the great philosopher, Thomas Hobbes (and others before him) wrote: "*primum vivere deinde philosophari*" (*first live, then engage in philosophy*). Likewise, Miguel de Cervantes relates a dialogue between two horses: "Are you a metaphysic?" asks one horse to the other. "No," replies its scrawny companion. "It is just that I haven't eaten for several days".

Even when their basic needs are satisfied, most people are quickly frustrated by the formidable complexity of the quest. It is like climbing mountains: the higher you reach, the larger appears the horizon (and the unknown beyond). As a consequence, many are content with easier goals, such as trying to accumulate as much wealth as possible (as though playing an endless game of Monopoly), and then bequeathing it when they die to somebody, who may well play the opposite game, and set about wasting it inanely.

However, the quest for knowledge has been an enduring and noble component of human endeavour over several millennia. In this book we try to provide an insight for the non-specialist into the science of physics, which is perhaps the most fundamental of the scientific disciplines, although we rebuff the extreme position, allegedly propounded by Lord Rutherford, that all science is either physics or stamp collecting. We have attempted to make our presentation as simple as possible, but not oversimplified,[2] lest the beauty of the physical concepts and arguments be lost.

So what is the meaning of life, the universe, and everything? We are not presumptuous enough to believe we can answer such a question, when millions of others, since time began, have already tried. However, we suggest that any attempt to understand the *purpose* of the universe must begin with a basic knowledge of what the universe *is*, and how it works.

1.3 Do Horses Exist?

To recapitulate the contents of the previous Section, our goal in this book is to try to learn as much as possible about the world in which we happen

---

[2] *Everything should be made as simple as possible, but not simpler:* usually attributed to Albert Einstein.

to exist, and to do so in a scientific way, i.e. using only our own rationality. From its very beginning, our task is formidable since, as we shall now see, even the notion itself of *existence* is not totally clear.

At first a debate about the meaning of existence might appear futile, but Segismundo's story warns us that it is not. Our dreams are sometimes so vivid that only when we wake do we realize that they were but dreams. Actors in a movie or theatrical play must immerse themselves totally in their characters, else they lose authenticity. For the duration of the performance, they give up their own personalities, and cry or laugh, suffer or rejoice as the characters they represent would presumably do, in order to interpret their roles convincingly. A suspension of disbelief by the audience is avidly sought by the director, and can be quickly shattered by one out-of-character line or anachronistic prop.

Also, *reality* may at times be so confusing, that we can hardly believe our eyes and recognize what is indeed *real*. The ancient Greeks remarked, already 2500 years ago, that if we look at a stick standing slanted and half immersed in water, we see it as if it were broken at the water surface. In this case, the illusion, illustrated in Fig. 1.1, is caused by a well-known physical effect, i.e. the refraction[3] of light rays at the water's surface. In other cases, such as those in Figs. 1.2, 1.3 and 1.4, it is our brain that constructs the illusion.

In Fig. 1.2, the tone of the red colour appears to be darker in the right-hand half of the diagram. This is known as the Bezold Effect, and is produced by the different coloured backgrounds in the two halves.

Figure 1.3 displays an image of a blivet, or "devil's fork". The image represents a physical impossibility, as such a shape cannot possibly exist.

This type of fantastical optical illusion has been developed into an art form by M.C. Escher in a series of beautifully detailed drawings, e.g. *Ascending and Descending* (1960) [2].

Other optical illusions can be produced simply by the image being taken from an unusual viewpoint. Figure 1.4 is a photograph of the so-called "super moon" taken at Bondi, Sydney, on November 15th, 2016.[4] The apparent large size of the moon is a consequence of the photograph being shot with a 600 mm lens. The moon-watchers in the foreground are actually located hundreds of metres from the camera. When we look at the image, our brains assume the foreground is much closer than this, and overestimate the size of the moon accordingly.

---

[3]Refraction is the bending of a ray (or wave) of light, heat or sound as it passes obliquely through the interface between one medium and another or through a medium of varying density.

[4]The moon was actually at its closest to the earth in 68 years on November 14th, but overcast conditions in Sydney prevented it being photographed on that day.

Fig. 1.1 A stick immersed in water appears broken at the water's surface due to refraction (i.e. bending) of the light as it crosses the boundary between water and air

Fig. 1.2 Bezold effect: Optical illusion produced by the contrast between two different coloured backgrounds. Image: Public Domain (https://en.wikipedia.org/wiki/File:Bezold_Effect.svg (accessed 2020/5/8))

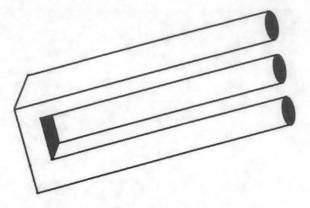

Fig. 1.3   Image of a fantastical figure, a Blivet or devil's fork

Fig. 1.4   Photograph of "super moon" at Bondi, Sydney, in November, 2016. The photograph was taken with a 600 mm lens. *Image courtesy* Janie Barrett/Fairfax Media

It is common, but incorrect, to assume that the telephoto lens is the source of this "distortion". If the photograph had been taken with a normal 50 mm lens, and the central part of the image cropped and enlarged, the effect would have been the same. Our brains have evolved to make sense of our environment, and are accustomed to an angle of view provided by our eyes equivalent to that of a 50 mm camera lens, which is the reason this lens is the default for everyday photography.

Nowadays, optical illusions are considered nothing more than amusing curiosities. The ancient Greeks, however, took them as proof of the deceitful nature of our senses. In his famous *Cave Allegory* [3], Plato compares men to slaves enchained in a dark cave with a blank wall before them. Other beings come and go, but the slaves cannot see them. The only evidence of their presence is their shadows, projected onto the wall by a fire behind them. Most slaves believe that the shadows they see are indeed the reality, but the philosophers among them try to reconstruct and ascertain what is actually going on.

Nor is it only our eyes that cannot be trusted. Our reason can also lead us astray, as was illustrated by the Greek, Zeno of Elea (ca 490–430 BCE), in a series of intriguing paradoxes. In his most famous example, he assumes that Achilles has a footrace with a tortoise, to which he gives a head start of 100 m. During the time it takes Achilles to cover the 100 m, the much slower tortoise advances a distance of (say) 10 m. Likewise while Achilles covers those extra 10 m, the tortoise keeps going for one more meter, and so on. Since the process goes on ad infinitum, Achilles will never be able to reach the tortoise because of the infinite number of steps required. The resolution of this paradox requires a realisation that the sum of an infinite number of finite quantities can actually be finite. (A simple example of such a summation is given in Appendix 1.1)

Another troubling issue for the Greeks was the ephemeral nature of life. If we see a horse, can we really say that it exists? After all, a few years ago it did not exist and in another few years it will have died, and passed back into oblivion. Can such a temporary condition really be called *existence*? Shouldn't we rather say that what exists is not the individual horse, but rather the *form* or *idea* of a horse, i.e. the *equinity*? Even if horses did not exist in the past, or will not exist in the future, surely the *idea* of a horse must have always existed.

Nowadays we know that not only appearances can be deceptive, but also that most of the world is completely beyond the grasp of our five senses. Even without entering into the most recent advances, such as dark matter and dark energy (which will be touched upon in later chapters), we all know that our senses of vision and hearing are both very limited, being constrained to the narrow range of frequencies (or wavelengths) of light and sound that our eyes and ears can capture. (These limits are detailed in Appendix 1.2.) Curiously enough, science warns us that the inadequacy of our senses is much deeper than ever anticipated by the ancient Greeks, but also it provides us with the tools to interpret the reality beyond.

For what concerns the reality of our ideas and concepts, Darwin's Theory of Evolution teaches us that, contrary to the dogma of the great Swedish biologist, Carl Linnaeus (1707–1778), species are in a continuous state of evolution themselves, so that an immutable idea of equinity does not exist: both horses and equinity have only an ephemeral existence. Science has helped us go one step further, beyond the experience of our senses.

But even with the full backing of science, i.e. when we arrive at all the conclusions that present-day science can provide to us, can we really state that what we have found corresponds to the full reality, i.e. to what really exists? Of course not, but, as we will learn in the next Section, we must be satisfied with that knowledge.

1.4   Is There Only One Truth?

In the previous Sections we have tried to define the goal of our book, i.e. to learn as much as possible about the *Universe* in which we live, relying both on what we perceive inside us (i.e. *a priori*) and on what we discover from the outside (i.e. *a posteriori*). In the process we must keep in mind that both our reason and our senses can at times deceive us. Hence, we must find criteria to guide us in our path and hopefully keep us on the right track.

Essentially we can identify two such criteria: *consistency* and *economy in the choice of assumptions*. We leave the latter for later discussion (in Chap. 3) and concentrate here on the former. It might appear that the need for consistency, i.e. of avoiding direct or indirect contradictions, is so obvious that no further discussion is required. However, in normal worldly affairs this is certainly not the case. An analysis of the speeches of most politicians shows that they often indulge in promises that are, at least in part, mutually conflicting, in order to appease as many voters as possible. More relevant for us here has been the proliferation in past centuries, and in totally different cultural contexts, of philosophical theories, which aimed at the justification of a *double truth*, e.g. a *scientific truth* and a *theological truth*, sometimes apparently in mutual contradiction, yet both considered to be valid in their own domains.

In order to understand what appears to be a contradiction in terms, we first must define what we call *truth*. If we mean that truth is what exists in an absolute sense, independently of our observations and immune from any possible contradictions, the question arises of how we can reach it. Indeed, there may be various alternative possibilities to achieve this, in which case we might arrive at multiple interpretations of the truth, each of them valid in its own context. It may seem astonishing that such apparent contradictions can take place in the realm of science. However, an eloquent demonstration of this is the centuries-long dispute about the nature of light.

Even in antiquity the ancient Greeks had wondered about this issue, but for brevity let us jump directly to Newton's Corpuscular Theory of Light (1704), in which a light ray was considered to comprise a beam of particles. Newton's proposal was generally accepted as true until Thomas Young in 1803 carried out an experiment that involved the diffraction[5] of light from two slits, thereby proving that light must have a wave-like nature. This was consolidated in the second half of the 19th Century, when James Clerk Maxwell (and others) interpreted light as a form of electromagnetic radiation. At this time, Newton's corpuscular theory was totally discarded as a somewhat quaint relic of a bygone era.

This rejection, however, turned out to be short-lived. The triumph a few decades later of Quantum Mechanics (QM) revealed that the corpuscular and wave characters of light are actually two different facets of the real nature of light, rather like the two sides of a coin. Both are correct, but they are incomplete and complementary, since both must be considered for a complete description of the observed phenomenology. This duality of QM has been extended from light to matter, as particles (electrons, protons, etc.) have been found in certain circumstances to exhibit a wave-like behaviour (see Chap. 5).

So, as we have seen, both the corpuscular and wave natures of light were considered at various times to represent the truth, but in effect they were both only provisional interpretations of the reality. Likewise, whatever we believe to be true in our time might in turn become outdated, if and when a higher-level "truth" is discovered. In Part 3 of this book, we will argue that some of the current discrepancies between otherwise quite successful theories might be due to their incompleteness, and we will search for hints that might allow us to look ahead in our quest.

At this point it might be interesting to ask what has happened to multiple truth doctrines and whether they are still acceptable. From a scientific point of view, the double truth doctrine helped scientists like Copernicus and his followers avoid being burnt at the stake. They claimed with sincerity that two conflicting truths could coexist, since they belonged to totally different domains: reason and religion. Others, such as Isaac Newton, held unorthodox religious views that they kept largely to themselves to avoid the consequences of the Blasphemy Act of 1697, which could have seen them stripped of all property and even sentenced to death.

---

[5]Light passing through two slits produces a pattern of light and dark bands, caused by interference between the two components of the beam that passed through each of the slits. See Chap. 5.

Outside the scientific domain, the double truth doctrine can be found in certain religions. In Buddhism, Buddha's teaching of the *dharma* is based on two truths, one purely of worldly conventions, while the other is the ultimate truth. Within the field of science, however, the double truth doctrine is no longer acceptable, and any conflict of ideas is regarded as an indication that our knowledge is incomplete, and more work is required.

1.5   An Oddly Named Asteroid

In this age of space exploration, we are all familiar with the nature of the Solar System, which comprises the sun and a number of planets orbiting with their satellites at increasing distances about this central star. Besides the planets, there are millions of other astronomical objects orbiting around the sun. Lying mainly between Mars and Jupiter are the asteroids, which are thought to be the shattered remnants of bodies in the young solar nebula that never grew large enough to become planets. Some of them are however rather large (up to 1000 km in diameter). Although they are not as well-known as the planets, their importance is now being recognised, a consequence of a growing fear that one of them at some time may collide with the earth. The United Nations has declared June 30th to be International Asteroid Day, in an attempt to bring them to the attention of the general population.

An example of an asteroid is shown in Fig. 1.5. Asteroid 243 Ida has an average diameter of 31 km, but is still large enough to have its own small moon, Dactyl, which is just 1.4 km in diameter.

What is relevant for our story here is Asteroid 11,059, which bears the name *Nulliusinverba*. This may appear a rather strange choice of name, until one realizes that it is also the motto of the British Royal Society. So why does a phrase owing its origin to a verse by the Latin poet, Horace, assume such importance in science that it is enshrined in the names of celestial bodies and mottos?

In English, the phrase means "on the word of no one" or "take nobody's word for it". The Royal Society website [4] explains its motto as an expression of the determination of Fellows to withstand the domination of authority and to verify all statements by an appeal to facts determined by experiment. It is an encapsulation of what separates (or in an ideal world should separate) science from other forms of human intellectual activity. That such an approach to science has not always been the norm is reason enough to remind us of its importance by a motto, or a name.

Fig. 1.5 Asteroid 243 Ida and its moon, Dactyl, which is shown to the right of the asteroid. *Image courtesy* NASA/JPL-Caltech (https://www.jpl.nasa.gov/spaceimages/details.php?id=PIA00136 (accessed 2020/9/1))

Let us turn for a moment to the history of philosophy and, in particular, to the legacy of the Greek philosopher, Aristotle (384–322 BCE), to fully appreciate the importance of Horace's message. Aristotle is the best known of all western philosophers, and certainly the one who has had the most influence upon his successors. His output was truly phenomenal and eclectic in character, with contributions to all the disciplines of his contemporary world, including logic, metaphysics, mathematics, physics, biology, botany, ethics, politics, agriculture, medicine, dance and theatre.

However, as a consequence of his success and fame, for about two thousand years any further development of philosophy was blocked, since it was much easier (and commonly acceptable) to search for the answer to any question in his books, rather than to research it independently. In other words, science, together with all other branches of philosophy, had ended up becoming the art of interpreting Aristotle's writings, rather than exploring new ideas. No innovation or progress was desired or even allowed.

In this scenario, the independent spirit of Horace stands out as a solitary opposing voice and illumines the path to follow in the pursuit of scientific progress. Eventually, of course, the individualistic streak in humanity rebelled against this subservient attitude and modern science was born, most notably with Galileo (1564–1642 CE) and his followers, who stressed the importance of observation and experiment in arriving at scientific truth.

Lest we become too smug and dismissive of eighty generations of past philosophers, who willingly placed their own ideas subservient to those of the demigod Aristotle, it is appropriate here to examine the current situation in the practice of science. It might be expected that an open-minded examination of new ideas, with no resort to prejudice, would prevail, in accord with the spirit of *nullius in verba*. However, the history of science abounds with examples of scientists who have had their work dismissed for reasons other than lack of scientific merit. (See, e.g. Barber [5].)

Max Planck, who introduced the concept of quanta, or energy packets, into physics (see Chap. 5) reported on how a paper containing his ideas was received by his university professors: *None of my professors at the University had any understanding for its contents. I found no interest, let alone approval, even among the very physicists who were closely connected with the topic. Helmholtz probably did not even read my paper at all. Kirchhoff expressly disapproved … I did not succeed in reaching Clausius. He did not answer my letters, and I did not find him at home when I tried to see him in person at Bonn. I carried on a correspondence with Carl Neumann, of Leipzig, but it remained totally fruitless* [6].

Mendel's ground-breaking work on the inheritance of physical characteristics, which is regarded as the harbinger of modern genetics, was ignored for thirty-five years, and only resurfaced in 1900 when Correns, de Vries and von Tschermak rediscovered his laws of inheritance. Mendel's work suffered from three damning faults: it ran counter to the prevailing wisdom on the inheritance of biological characteristics; it made use of mathematics, unheard of at that time in biology, and its author was only an unknown monk from Brunn, and not a member of the scientific establishment.

Others, now household names in the world of science, suffered similar discrimination. Ohm's work on electricity was ignored, partly because he was of insufficient professional standing, being only a mathematics teacher at a Jesuit High School in Cologne. Pasteur's discovery of germs and their role in infection was spurned because he had no medical qualifications. Oliver Heaviside's *cri de coeur*: "*even men who are not Cambridge mathematicians deserve justice*" was a consequence of his work in mathematical physics being ignored for twenty-five years.

It is interesting to examine some of the motivations of those who impeded the acceptance of scientific works that later turned out to be seminal. For instance, some scientists bore prejudices based on their own approach to science, and were unwilling to credit that others with a different background or *modus operandi* could produce anything of substance. This tendency was exacerbated when the work arriving on their desk was from a little-known author, and challenged established scientific beliefs.

Resistance to the heliocentric model for the solar system of Copernicus was opposed, not only by the Catholic Church, but also by astronomer-scientists such as Tycho Brahe. Lord Kelvin liked to make mechanical models to facilitate his understanding, and had trouble with abstract concepts, such as the electromagnetic theory of Maxwell [7]. As we have seen above in discussing Mendel's fate, Kelvin was not alone in his opposition to a mathematical approach to science. Indeed, the lack of success by mathematically inclined biologists in getting their work published in traditional journals led them to found their own journal: *Biometrika*.

Philosopher, Francis Bacon, the father of *Empiricism*, was influential in his opposition to theoretical science. However, he and his followers were dismissed as "bird-watchers" by the more mathematically inclined, some of whom carried their own prejudices so far as to scorn the discoveries on electromagnetism by Michael Faraday, one of the greatest of all experimental physicists [8]. The religious beliefs of some scientists led them to oppose specific areas of science, such as evolution and geology.

Generally speaking, it is the younger scientists who are more receptive to radical ideas than their older colleagues. As Max Planck once stated: *A new scientific truth does not triumph by convincing its opponents and making them see the light, but rather because its opponents eventually die, and a new generation grows up that is familiar with it.*

It is now over half a century since Barber's paper [5] on scientific bias appeared, and one might think that attitudes had by now surely changed. It is probably true that the practice of "open-mindedness" is more common in science than in most other professions. At the heart of science is the peer review process, whereby research submitted for publication in a science journal is subjected to a critique by several anonymous "experts" in the same field. More than one million research papers are published each year, and reviewers are usually expected to provide their services for free. Rejection rates are high for the most prestigious journals. Modern computer databases keep track of how often individual papers, authors and journals are cited by other researchers, and these data impact on scientists' promotion prospects and chances of getting financial support for their projects.

Science has become a very competitive industry, and scientists being only human, it should be no surprise that some of them resort to "grey" practices to promote their own ideas and work at the expense of their rivals. Cases of outright fraud are however rare. In the long run however, the replication of experiments and observations in different laboratories by different researchers usually sorts out what is correct from what is not, even if sometimes it may take a rather long time.[6]

1.6   The Limits of Science

In a now lost poem, a sequel to *The Odyssey*, by Greek poet, Homer (born sometime between the 12th and 8th Centuries BCE), Ulysses managed to convince his sailors to attempt what was then considered the ultimate quest, i.e. to explore the uncharted waters of the Atlantic. This adventure involved passing through the Strait of Gibraltar and past the columns set there by the demigod Hercules, which were inscribed with the phrase: *non plus ultra* (no more beyond), to warn seafarers not to venture beyond.

After countless days of navigation, the fearless men finally spotted land, but their joy soon ended in despair when a sudden storm and whirlpool wrecked their ship. The interpretation of this tragic outcome by Dante (and Christians) was that the land they arrived at was the mountain of Purgatory, which was of necessity forbidden to all living men. Ulysses had perished in a quest for forbidden knowledge.

Likewise many other limits to human knowledge have been assumed at various times. In the *Bible*, Adam and Eve are punished with the loss of the Earthly Paradise for having tasted the fruit of the Tree of Knowledge. When the Babylonians displayed haughtiness by attempting to build a tower extending to heaven, God thwarted their plans by confusing the language of the builders so that they could no longer communicate with each other. As a final example, cartographers in medieval times decorated their maps with pictures of monsters, and warnings such as " *Hic sunt leones*" (Here be lions) and " *Hic sunt dracones*" (Here be dragons), to dissuade incautious visitors. An example is shown in Fig. 1.6. It would be a brave mariner indeed who would use such a map to venture too far from shore.

In more recent times science has shaken many dearly held beliefs: the Copernican system has replaced the revered Ptolemaic model of the solar system, and Darwinism has dismantled faith in the immutability of animal species and in the special place of humankind in the scheme of creation. As a reaction, many followers of religion (often using tools and means provided

---

[6]A classic example of this process is the discovery of gravitational waves in 2016 (see Chap. 7). A whole field of science was built on attempting to verify an erroneous experiment by Joseph Weber in 1972.

Fig. 1.6 A portion of a marine map of Scandinavia, drawn up by Olaus Magnus at Venice in 1539, showing some of the dangers likely to befall unwary seafarers who venture too far from the land. Image: public domain (https://commons.wikimedia. org/wiki/File:Carta_Marina.jpeg (accessed 2020/8/30))

by science) have suggested that scientists can sometimes be wrong, which is undoubtedly true, and that science has bounds which only religion can cross.

Let us leave aside any controversy between science and religion, and reiterate that in this book we wish to explore, from a purely scientific point of view, the vexed question of the limits of science. It is a timely question, since just two or three decades ago a number of scientists were confident of being close to a TOE–a *Theory of Everything*. By this they meant that all the forces of nature could be described by one overarching theory. Such a situation would have killed and buried physics as an active field of research. As we shall see later in this book, a TOE has not come to pass, and may actually be an impossible quest.

The aim of physics is to explain reality, but as we have seen in the earlier Sections of this Chapter, it is not really possible to say with any certainty what reality is. One might simplistically say it is what we perceive with our senses. However, our senses are very limited, and science has progressed largely by developing instruments and techniques to extend their powers. Microscopes and telescopes aid our eyes to see very small and very distant objects, but how do we know that these objects really exist?

Werner Heisenberg, one of the founders of Quantum Mechanics, recognized the important role that the observer plays in our notion of reality: *We have to remember that what we observe is not nature in itself, but nature exposed to our method of questioning* [9]. Our view of reality is myopic, limited by our instrumentation, and quite possibly distorted, like that of a goldfish viewing the world outside of its bowl through the curved glass. As poet William Blake wrote: *If the doors of perception were cleansed, everything would appear to man as it is, Infinite. For man has closed himself up, till he sees all things thro' narrow chinks of his cavern* [10].

The belief is that external to us there exists a "true" reality, but there is no way to prove such an assertion. Would an alien, assuming one exists on another planet somewhere, see the same reality as we do?

As we will see in Part 2 of this book, the progress of physics from the turn of the 20th Century has been mainly to extend our "understanding"—a term that we clarify in the next Chapter—to the very small and to the very large and distant. However, our pictures of these two regions of reality are somehow contradictory, and problems arise when they overlap. Will future instrumentation enable this dilemma to be resolved, or will some questions always remain unanswerable? One may regard science as a methodology for testing out various hypotheses that we, as human observers, hold about the physical world. We may strive towards truth, but the more questions we answer, the more new ones that surface and require an answer.

Another issue to be considered is whether society will continue to support research into areas such as these, which, although indulging humankind's curiosity into the origin of the world and of life, produce few practical outcomes commensurate with its huge costs. In other areas, e.g. modification of the human genome, human ethics have placed limits on research, when the knowledge obtained would likely produce undesirable social consequences.

The same cost restraints do not apply to theoretical physics, but as we will discover in Chap. 4, other issues become relevant. We have already seen in Sect. 1.3 how Zeno's paradox of Achilles and the tortoise puzzled philosophers for centuries. Eventually this enigma was resolved, but others remain. The field of mathematics was shocked by the incompleteness theorems (see Chap. 4) proved almost a hundred years ago by the Austrian mathematician, Kurt Gödel (1906–1978), which showed that some propositions in mathematics are unprovable. As physics is based on mathematics, surely similar limits to knowledge must apply also in physics.

In considering questions such as we have raised here, it is important that the debate is not left entirely in the hands of scientists. Science concerns us all, since it has brought countless benefits to the whole of humanity. It has also brought its shares of woes. Science can be a frightening toy in the hands of the amoral or malevolent. For most people what is really relevant to their everyday lives is the use of technologies, such as those required by smart phones, computers, navigators, cars and planes. Behind all such inventions there is a huge amount of science, and surely it makes no sense at all to take advantage of the inventions and innovations, while not believing in the science that makes them possible. A keen and well-informed awareness of science is an asset for everybody, and not just for scientists.

References

1. Shakespeare W, Macbeth, Act V, Scene V
2. https://mcescher.com/lw-435/#iLightbox[postimages]/0. Accessed 8 May 2020
3. Plato, Republic (514a–520a), written as a dialogue between Plato's brother, Glaucon, and his mentor Socrates, narrated by the latter
4. https://royalsociety.org/about-us/history/. Accessed 9 May 2020
5. Barber B (1961) Resistance by scientists to scientific discovery. Science v134:596–602
6. Planck M, Scientific Autobiography F (1949) Gaynor, transl. Philosophical Library, New York, p p18
7. Thompson SP (1910) The life of William Thomson: Baron Kelvin of Largs. Macmillan, London
8. Gillispie CC (1960) The edge of objectivity. Princeton Univ. Press, Princeton, NJ
9. Heisenberg W (1958) Physics and philosophy: the revolution in modern science. Lectures delivered at University of St. Andrews, Scotland, Winter 1955–56
10. Blake W, The marriage of heaven and hell

2

The Meaning of Understanding

2.1   The Origin of Intelligence

Mal Looked at the Water Then at Each of the People in Turn, and They Waited.
"I Have a Picture."
He Freed a Hand and Put It Flat on His Head as if Confining the Images
That Flickered There.
"Mal is not Tall But Clinging to his Mother's Back. There is More Water Not
Only Here But Along the Trail Where We Came. A Man is Wise. He Makes Men
Take a Tree That Has Fallen and —".
His Eyes Deep in Their Hollows Turned to the People Imploring Them to Share
a Picture with Him. He Coughed Again, Softly. The Old Woman Carefully Lifted
Her Burden.
At Last Ha Spoke.
"I Do not See This Picture."
The Old Man Sighed and Took His Hand Away from His Head.
"Find a Tree That Has Fallen."
The Inheritors: William Golding [1]

In this short excerpt from the *"The Inheritors"* by Nobel Prize winning author, William Golding, we encounter a Neanderthal family confronted with a serious problem: a fallen tree that served as their bridge across a river has been swept away. Mal, an old man, recollects a similar situation from the time when he was an infant, clinging to his mother's back. He tries to convey this memory to his family, but is unsuccessful. He proposes that they find another fallen tree, and drag it to the crossing. This is a major innovation:

to the other family members, the original fallen tree, like the sun, the moon and the river, had always been there.

Golding provides us here with a glimpse of the possible origins of rational thought. Is this how it began, with the sharing of experiences and memories? Did intelligence evolve from this exchange of mental images?

Archaeological research and findings from the Human Genome Project suggest that modern cognition and behaviour had developed in sub-Saharan Africa by at least 50,000 BP.[1] The last glacial period, or Ice Age, extended from 110,000 to 12,000 years ago. During this period, the environment changed markedly, with the sea level falling and rising by many metres. The changing environmental conditions during this period may have provided an evolutionary advantage to those primates with a higher intelligence, and influenced the evolution of modern humans [2]. The spread of *Homo sapiens* from Africa to other continents appears to have occurred in several waves during this period, and to have been facilitated by low sea levels arising from large masses of water being tied up in glaciers. Genetic evidence suggests that the first exodus occurred about 70,000 years ago, and followed the coastline of Asia to Australia, which was reached at least 50,000 years ago. Subsequent migrations led to the populations of Central Asia, Europe and the Americas [3, 4].

Today intelligence appears to be dependent on a multiplicity of genes, as well as the environment, so it is likely that it is a characteristic that evolved over time, rather than from a single spontaneous genetic mutation [5].

Before discussing intelligence, we need to be clear what exactly we are talking about. Howard Gardner in his 1983 book, *Frames of Mind: The Theory of Multiple Intelligences*, lists nine different types of intelligence [6]. Some of these involve musical ability and hand–eye coordination, which some might regard as aptitudes, rather than intelligence. In our book we are interested primarily in Gardner's category: *Logical-Mathematical Intelligence*, which deals with logic, abstractions, reasoning, numbers and critical thinking.

As intelligence evolved, humans (or *Homo sapiens*, as *Homo Neanderthalensis* had by now either lost their identity by interbreeding with their newly arrived cousins, or become extinct) turned their skills to survival. They learned how better to protect themselves and their families from the elements and wild beasts, and how to gather enough food on an almost regular basis. A higher level of intelligence improved their chances of survival.

If intelligence offered humanity an increased prospect of survival in a hostile world, as a by-product it provided them with an awareness of the

---

[1]Before the present.

world around them. They could not fail to notice and exploit the cycles of nature. Day changed into night, then back to day again. The seasons succeeded one another inexorably through the course of a year. The exception appeared to be humans themselves, who aged or died through misadventure or disease and were lost forever. How could this be? Perhaps death was not the end, but just a doorway into a different cycle. Perhaps we do not entirely die, and some part of our essence remains.

Of course, what we have described above is conjecture. However, we do know that primitive people were already interested in their surroundings. Figure 2.1 displays rock art from the Lascaux Caves in France, depicting aurochs, horses and deer. It is estimated to be up to 20,000 years old.

The Chauvet-Pont-d'Arc Cave in southern France contains rock paintings dating to about 30,000 years ago of a variety of animals. Australian Aboriginal rock art has been found in Western Australia's Pilbara region and the Olary district of South Australia, and the oldest is about 35,000 years old. It depicts extinct megafauna, e.g. *Genyornis newtoni*, a flightless bird over two metres tall, and *Thylacoleo*, a formidable marsupial carnivore.

Since humans were alive, they interpreted everything around them as also being animated. Unable to understand natural bodies and their behaviour, they attributed to them a divine nature. Their description of the world was

Fig. 2.1 Cave paintings of aurochs, horses and deer from Lascaux caves (Montignac, Dordogne, France). Image, courtesy of Prof saxx (Creative Commons: https://commons.wikimedia.org/wiki/File:Lascaux_painting.jpg (accessed 2020/8/11))

mythological, i.e. intertwined with symbols and legends. To answer the question of the world's origin, some primordial form of cosmogony was assumed: the shapeless chaos of Greek mythology, the waters of the Bible and of Mesopotamian and Egyptian mythologies, the qi of the Chinese tradition. Gods were also conceived, like any other form of life, as the result of some inexplicable and sudden event.

In the dreamtime of the Australian aboriginals, the earth began as a bare plain, with no life or death. Then the eternal ancestors rose and found half-formed shapeless human beings with no limbs or features, created from animals and plants. The ancestors carved out heads, arms and feet, and finished the creation process. They then went back to sleep, leaving traces of their presence in what are today sacred sites. To the aboriginal peoples, the dreamtime is not in the past, it is eternal.

Both the Chinese and the Hindu traditions conceive the evolution of the universe as a cyclic process. The Hindu cosmology even quantifies the duration of each cycle in terms of billions of years. At the end of each cycle the universe is destroyed by fire, and then after an interlude, it is created again.

It is instructive to realize that human beings have long been awestruck by the mysteries of nature. One might even argue that it is this yearning for knowledge that defines the species, *Homo sapiens*. However, their intuitive approach was very remote from the formal mindset of modern science. In the next Section, we will consider the formal logic, or systematised way of thinking, that underpins modern science.

2.2   Logic, and Its Role in Science

Science, and the mathematics that underpins it, is derived from formal logic, which is a systematised way of valid reasoning. Formal logics were developed in China, India and Greece. The roots of logic probably began with the development of language. In modern humans both of these activities appear to occur largely in the left hemisphere of the brain. There is no consensus on when spoken language first appeared, with estimates varying widely. Because there was apparently no way to tackle this question in a meaningful, i.e. scientific, manner, the Linguistic Society of Paris in 1866 banned any existing or future debates on this subject. The prohibition remained influential for over a century [7].

It is far beyond the scope of this book to delve into the history and nature of formal logic. After all, it is a subject that has occupied philosophers for centuries. However, a few points are worth considering to set the background for the material we wish to cover later. In the West, Aristotle, whose work inspired Euclid, Archimedes, Apollonius and many others, pursued a bottom-up approach to rational thought. They began from simple concepts that could be accepted readily (axioms), and deduced more complicated consequences by following strict rules of reasoning. The geometry of Euclid that we all learned in high school is probably the best-known example of this approach.

A phrase such as the "strict rules of reasoning" might give the impression that these rules are universally understood and applied in all daily human intercourse. This is certainly not the case. As an example of logic, consider the two statements: "no desire is voluntary" and "some beliefs are desires". The logical consequence of these two statements is the conclusion: "some beliefs are not voluntary". The underlying logic of this thought process can be extracted and formalised by writing it in the form:

"If no Q is R and some Ps are Qs, then some Ps are not R."

Here Q is shorthand for "desire", R for "voluntary", and P for "belief". The same logic is applicable to countless other examples. For instance, if Q = "flyer", R = "pig", and P = "bird", the above logical statement leads to: "If no flyers are pigs, and some birds are flyers, then some birds are not pigs." It might be tempting to cry out: "but surely *no* birds are pigs."[2] While undoubtedly true, this stronger conclusion cannot be deduced from our two premises. Such is the uncompromising nature of formal logic.

So confident was Aristotle in the reliability of this method that, when he applied it to the physical world, he seldom bothered to check whether his inferences were in accord with actual reality. For instance, he asserted that a moving body in a vacuum came to rest immediately after the force that had induced the motion was removed. In air, he believed the motion persisted because the surrounding air sustained the force for a while, until it ultimately dissipated. This "fact" was accepted by scholars until finally refuted by Galileo, almost two millennia later. It is the combination of logic and observation, beginning with the work of Galileo, that is the characteristic of modern science.

Before leaving this topic, there are several questions that are worth posing, even if there is little hope of agreeing on the answers:

(1) Why does our mathematical logical approach explain the observed physical world so well?

---

[2] Strictly speaking, we should define what we mean by birds, pigs and flyers before introducing them into such a logical statement.

(2) Does logic, which is a product of our brains, have any existence outside of us?

(3) Assuming that other rational beings exist in the universe, would they necessarily develop the same system of logic that we have, or could they possibly find some other system equally capable of explaining their world?

Mathematics is often described as *a priori*, meaning that it exists independently of the outside world, in contradistinction to an *a posteriori* knowledge, which is obtained by empirical observation. This implies that mathematics would exist even if the world around us vanished. The brains of humans, and presumably other intelligent beings, harness this universal resource in the development of science. The term *Platonism* is used to describe this philosophical tenet. Eccentric 20th Century mathematician, Paul Erdös, often referred to "The Book", in which God keeps the most elegant proof of each mathematical theorem. He once said in a lecture: *you don't have to believe in God, but you should believe in The Book* [8].

Other philosophers deny that mathematics is *a priori* at all, claiming that it arose in the search for the best description of experience, and in that sense is no different from the other sciences. This viewpoint is known as *Empiricism*, and has been propounded by 20th Century philosophers, Willard Van Orman Quine and Hilary Putnam. A criticism of an empirical view of mathematics is that if mathematics is just as empirical as the other sciences, then its results are just as fallible as theirs.

The empiricist explanation opens the way for the evolution of logic and mathematics in the brains of early humans as a survival aid in the process of Darwinian natural selection, and gives insight into why mathematical logic works so well at describing the physical world. Of course, once developed, logic can be applied to any other abstract field not connected with survival. This is nothing new: our eyes did not evolve to read computer screens, but serve that purpose just as well.

We will re-examine these two alternative views of the nature of logic (and mathematics) in Part 3, after a review in Part 2 of what we have learned from the last century of progress in physics.

2.3   Pattern Recognition

Leaving aside these questions, which will no doubt occupy philosophers for another few centuries, it is worth considering whether mathematics and logic are the only approaches to a rational understanding of the universe. Formal logic plays little role in the thought processes of many, who rely on a more intuitive approach to their decision making. It has become customary to designate individuals as either "left-brained" or "right-brained", depending on whether they are analytical thinkers or intuitive. The differentiation is based on the fact that logical reasoning, along with language, seems to take place in the left cerebral hemisphere, while more holistic activities—art, music, etc.—take place in the right hemisphere.

This was always a doubtful distinction, as many people combine artistic and scientific aptitudes, and intuition—the sudden blinding flash of inspiration coming apparently from nowhere—is a valuable component of the scientific creative process (see Appendix 2.1). The story of Archimedes jumping up and racing naked down the street shouting "Eureka", after the idea for his famous Principle of Buoyancy occurred to him in his bathtub, is part of the folklore of science. Most research scientists have had their own Eureka moments, sometimes in a dream. For example, the German organic chemist, Friedrich August Kekulé (1829–1896), recounted that he had discovered the ring (or ouroboros[3]) shape of the benzene molecule during a day dream.

Returning to our point above, we query whether the bottom-up approach of logic is the only worthwhile path to the understanding of the workings of nature. In everyday life we have no difficulty in recognising a photograph of a friend. However, an ordinary computer, which can perform arithmetical calculations at a speed millions of times faster than a human, has difficulty carrying out this simple operation, which even an infant can readily achieve.

The reason lies in the complexity of our brain, which comprises billions of cells interconnected in a vast "*neural network*". Many processing operations take place in parallel, unlike the sequential processing of a normal computer. To reduce the facial recognition task into a series of steps for a computer to carry out sequentially is fraught with difficulties.

Most of the brain's parallel processing appears to take place subliminally, without any direction from the consciousness of its owner. From time to time this "subconscious" network throws the results of its cogitations up into the conscious component of our mind, and so we have Archimedes jumping out of his bath and shouting "Eureka", or a crossword enthusiast searching for yesterday's newspaper to fill in the answer to a clue that she had spent hours on, before giving up in frustration.

---

[3]An ancient symbol of a snake eating its own tail, often used in pop culture, e.g. for tattoos. It can be interpreted as a symbol for cyclic renewal, or an unending cycle of life, death, and rebirth. We encounter this symbol several times in this book.

Clearly logic is an important component of the brain's activities, but it is certainly not the only component, nor probably the most important for our daily survival. Pattern recognition, which enables us to separate objects into particular classes, even if we have never seen them before, is a vital part of human learning. We all know a tree when we see one, and do not attempt to teach it to heel, in the belief that it is a dog. The extraction of patterns has played a vital role in our survival. It is a facility that is trained into the minds of infants by their earliest childhood experiences. Pattern Recognition is currently being incorporated into computers in research into "machine learning", with the aim of further developing Artificial Intelligence.

2.4   Complexity

The bottom-up logical approach has been the traditional *modus operandi* of physics. For instance, the laws of interaction of particles were proposed by Newton and others, and the result was the science of classical mechanics. The orbits of planets, and the paths of rockets, have been deduced from these laws by the use of mathematics. Different laws were formulated for the interaction of high-velocity bodies by Einstein, leading to relativistic mechanics, and by Heisenberg, Schrödinger and others for sub-atomic particles, leading to Quantum Mechanics. Deductions from these laws have led to the prediction of phenomena that have been observed experimentally.

Difficulties arise when attempts are made to apply these physical laws to scenarios with large numbers of interacting particles. It is not because anyone believes the laws do not work. Rather, it is that the mathematics of the problems becomes intractable. As an example, Newton's Theory of Gravity yields an exact analytical solution only for the case of *two* interacting bodies. One might imagine the situation of the earth revolving about the sun. The orbit of the earth can only be predicted analytically if we disregard the presence of the earth's moon, and of the other planets and their moons.

However, we know that during the Apollo missions, NASA predicted the paths of their spacecraft very precisely. How was this possible? Numerical methods have been developed which involve computing the effect on the rocket's trajectory of one body (the sun or the earth), and then refining the estimates obtained by repeating the calculations, including more and more "perturbations" from hitherto neglected gravitational sources (i.e. other planets and moons). Such a procedure is time-consuming, and only possible because of the development of fast modern computers.

Impressive as the space program undoubtedly was, the score or so of inter-acting bodies involved in these calculations is negligible compared with the approximately 10^{22} molecules[4] in a jar of air. Numerical methods cannot provide a way to handle the interactions of this number of molecules, and physicists have been forced to resort to a statistical approach.

In "Statistical Mechanics", the behaviour of molecules is only considered *en masse*. The bulk properties of matter are studied, and concepts such as temperature and pressure introduced, which arise from the average behaviour of large numbers of molecules. Temperature is related to the average energy of motion of the molecules, and pressure to their impact on the walls of the vessel containing them. New laws of physics are formulated which connect these bulk properties. For instance, increasing the pressure of a gas confined in a flask results in a proportional increase in the temperature of the gas.[5] This relationship was first discovered by Joseph Louis Gay-Lussac in 1809.

There is a difference between these "statistical" laws and the more funda-mental laws describing the interaction of particles. The former may in principle be derivable from the latter, and as such may not be considered to be basic laws of physics at all. Although this may be true in some cases, in most scenarios such derivations are not possible because of the complexity of the interactions and the huge numbers of particles involved.

The science of Thermodynamics was developed in the 19th Century, and was motivated by a desire to increase the power and efficiency of steam engines. Its laws do not relate to interactions between individual particles, but rather involve higher level concepts, such as heat, temperature and entropy. The term "entropy" has been coined for a measure of the disorder of a system. The universe is analogous to a child's playroom, where the toys start out in the morning neatly arranged on shelves and in boxes, but at the end of the day have become strewn randomly over every horizontal surface. Left to itself, nature tends to the state of maximum disorder. This tendency is expressed in what is known as "the Second Law of Thermodynamics", i.e. "entropy tends to a maximum." (A popular skit on this topic is discussed in Appendix 2.2.)

The Second Law is often stated in the alternative form: "heat cannot spon-taneously flow from a colder location to a hotter location." If we consider two flasks of gas, one at a higher temperature than the other, and connect them together with a tube, heat is gradually transferred from the hot flask to the cold one, stopping when the gas in both flasks is at the same temperature.

---

[4]Scientific notation for 1 followed by 22 zeroes.
[5]The temperature in this case is the *absolute temperature*, which is measured in the Kelvin scale. The absolute temperature is obtained by adding 273.15 to the temperature in degrees Celsius.

We never have a situation where heat flows in the other direction, thereby increasing the temperature differential between the gases in the two flasks.

If we consider this situation from a microscopic point of view, molecules in the hot flask are travelling on the average faster than those in the cold flask. The interconnecting tube enables the molecules to mix and collide with each other, with the result that on average the molecules of hot gas lose energy and those of cold gas gain energy, until the temperature difference between the two flasks vanishes. The Second Law states this principle formally, i.e. that we never expect the temperature difference between the flasks to increase. Such an action would be equivalent to replacing the disorder of completely mixed up gases with a situation where there are more hot molecules in one flask than in the other. The latter situation is less disordered than the first, a violation of the first statement of the Second Law.

Imagine now that we have a minimal amount of gas (very few molecules) distributed between the two interconnected flasks so that the temperatures in the two flasks are the same. It is quite possible in this case that random collisions might produce a situation whereby, for a while, more hot gas molecules are in one flask than the other, and we would have a temporary violation of the Second Law.

We can draw an analogy with the tossing of a coin. On the average we expect as many heads as tails when we toss an unbiased coin. If we toss the coin a million times, our expectation is that the number of heads will be within about 0.1% of the number of tails. However, if we only have three throws, the likelihood of all three producing the same result is reasonably high (1 in 4). The Second Law is similar, in that it is a statistical one; i.e., it applies very accurately when we have large numbers of molecules taking part in the collision processes. In a situation where there are approximately 10^{22} molecules in the two jars, the law is essentially exact.

Essentially exact, but not quite. We shall discuss the importance of this difference in the next Chapter, where we seek further insight into the nature of physical truth.

Let us return now to the main topic of this Section. Complexity is a relatively new field of study arising from a recognition that there are areas of science, that are too complex to be tackled by the conventional bottom-up methodology. A statistical approach is the only available way for some problems. An everyday example occurs in meteorology, where it is quite common to read a weather forecast along the lines that the chance of rain tomorrow is 60%, with a 30% chance of an afternoon thunderstorm. This may be frustrating if one is planning a picnic and would like more certainty about what to expect from the heavens. However, that is the best prediction possible at the present stage of development of meteorology.

The statistical nature of the predictions is not, however, the only property of a complex system. Another is the similarity of patterns within a given complex system at different levels of analysis or, as we shall see later, among different complex systems. Figure 2.2 depicts the vein structure in a leaf. The large, course veins and the finer ones show the same type of branching structure. This is a phenomenon observed widely in nature, e.g. in the large bays, smaller inlets and rock pools of a coastline. It has been reproduced using "fractals" generated by computer programs. Fractals are infinitely complex patterns that are self-similar across different scales or levels. They can be obtained by repeating a simple process over and over again in an ongoing loop.

Much publicity has been given to the very beautiful fractal patterns produced by Benoit Mandelbrot, an example of which is shown in Fig. 2.3. If the pattern in Fig. 2.3 is subdivided into ever smaller areas, one notices that the large scale basic structure is reproduced in all the subdivisions, in analogy with the naturally occurring structure of leaves and coastlines.

Not only can similar behaviours be found at different levels within a single complex system, but also some common patterns can be observed over and over again in completely unrelated contexts. Examples are the growth curves

Fig. 2.2 The veins in a leaf, showing a fractal-like pattern. Image courtesy of Curran Kelleher (https://www.flickr.com/photos/10604632@N02/922705627 under CC licence https://creativecommons.org/licenses/by/2.0/ (accessed 2020/5/10))

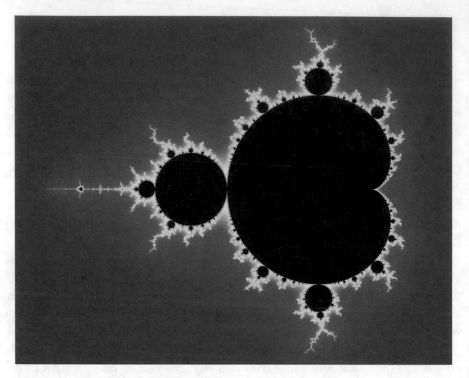

Fig. 2.3 Example of a Mandelbrot Set. The Mandelbrot Set is generated by successive repetitive applications of a simple mathematical formula. Image courtesy of Wolfgang Beyer (https://commons.wikimedia.org/wiki/File:Mandel_zoom_00_mandelbrot_set.jpg (accessed 2020/5/10))

associated with various types of tumours, crystals, prices (e.g. in the stock market), defects and infiltrations in materials, etc.

The same is also true for the growth of animals. Figure 2.4, reprinted from West et al. [9], shows the relationship between the mass as a function of time (i.e. the growth) for a wide variety of animal species. For example, superficially, if one were to compare the growth rates of a pig and a hen, they would look quite different. However, when plotted in dimensionless coordinates, as is the case in Fig. 2.4, a remarkable similarity emerges between the growth rates of all animal species.

Dimensionless units are used in physics, and in other sciences, to avoid dependence on arbitrary units, such as the metre and the kilogram, which have been specified by humans and have no particular physical significance. For instance, a person's mass may be specified as 77 kg, or 170 lb. In this case, his or her mass is being compared with that of a lump of metal lying in a Paris

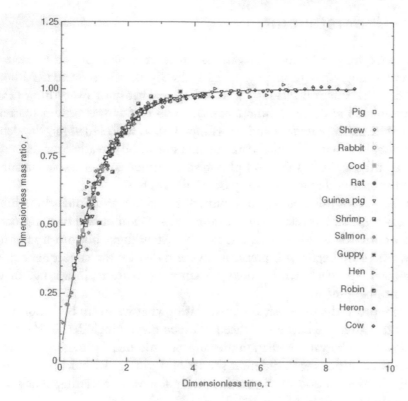

Fig. 2.4 A plot of the dimensionless mass vs. dimensionless time for a wide variety of species. Figure reprinted from ref [9]. by permission of Springer Nature, Copyright (2001)

vault.[6] By comparing the mass of an animal with some other more relevant measurement, perhaps the mass of the same animal at birth (or hatching), one can obtain a dimensionless value for the mass at any time after the animal's birth. Time may also be expressed in dimensionless units, by comparing the elapsed time with some other relevant unit of time (e.g. the average lifetime of individuals of that species). (For the dimensionless units of time and mass actually used in Fig. 2.4, the reader is referred to Ref. [9].)

What is striking in Fig. 2.4 is that when the appropriate choice is made, the growth curves of many species collapse onto a single graph. If we infer that this pattern is applicable to *all* animals, we have in Fig. 2.4 a law expressing the growth of *any* animal's mass as a function of time. This is a very far-reaching and powerful statement.

---

[6]This was true until 2018. The kg is now defined in terms of fundamental Physical constants (see Chap. 3).

2.5   Understanding

As we have seen in Chap. 1, at various times in history, physicists have felt that everything of note in physics had already been discovered. Although research work was still zealously proceeding, a Theory of Everything (TOE) was postulated and the common wisdom was that it was only a matter of time before all the *details* would be wrapped up as well. However, the demise of physics, like that of Mark Twain, turned out to be "greatly exaggerated": a TOE is still sought in vain, and physicists continue to seek explanations for new phenomena that are currently being discovered.

At the start of this chapter, we explored the origins of rational thought, the nature of logic and its relationship with physics. We discussed two approaches that are currently in use in the quest for understanding: a bottom-up methodology, where we begin with physical laws and deduce the consequences, and a top-down approach, which looks for common patterns in a range of very different phenomena.

Before proceeding further, we must clarify what we mean by "understanding", a term that we have introduced glibly and even included in the title of the Chapter. The various dictionaries are of little help, providing us with a wealth of definitions with different shades of meaning. Richard Feynman has added his contribution to the confusion by famously declaring that no one *understands* Quantum Mechanics.

For our purposes in this book, let us try to firm up our interpretation of "understanding" by comparing it with "knowledge". In the course of our lives we gain knowledge of many facts, which have been accrued from our observations, readings, and other sources of information. In general, these facts are collected one by one, as if they were isolated among themselves. However, when we have garnered enough of them, we may start to notice relationships between some of them. This is the beginning of what we mean by understanding: true understanding occurs when these relationships enable us to make predictions. In fact, predictability based on "understanding" is the essence of science.

Some would argue that true understanding comes only when physical laws are stated, from which mathematics can be used to deduce new facts, which are in turn verified by further observations. This is the bottom-up approach that we have discussed above, and whose power is demonstrated by the success of Newtonian Mechanics. However, as we shall see in Chap. 4, even this rigorous methodology may contain a fatal flaw embedded in its heart.

In Sect. 2.4, we saw that the less formal approach of pattern matching may provide us with a different kind of understanding. If Fig. 2.4 does indeed represent a true growth "law" for the animal kingdom, we can expect it to be valid also for animals (e.g. dogs, cats, lions, etc.) not included in Fig. 2.4. Otherwise Fig. 2.4 is little more than a collection of curious coincidences. As is always the case in science, further observation remains the final arbiter.

So, while it is true that the bottom-up approach lies at the core of physics, and will always remain the bedrock on which physical understanding is built, in some circumstances the complexity of nature forces physicists (and other scientists) to countenance a less formal, top-down, methodology. In the next chapter, we will explore some of these ideas further.

References

1. Golding W (1955) The inheritors. Faber and Faber, London
2. Calvin W, The ascent of mind: ice age climates and the evolution of intelligence (originally published by Bantam Books, 1991, 2000)
3. https://australianmuseum.net.au/the-spread-of-people-to-australia. Accessed 10 May 2020
4. Wells S (2000) The journey of man: a genetic Odyssey. Penguin
5. Sternberg RJ, nd Grigorenko E (eds) (1997) Intelligence, heredity and environment. Cambridge University Press, Cambridge
6. Gardner H (1983) Frames of mind: the theory of multiple intelligences. Basic Books
7. Stam JH (1976) Inquiries into the origins of language. Harper and Row, New York, p 255
8. Hoffman P (1999) The man who loved only numbers: the story of Paul Erdos and the search for mathematical truth. Hachette Books
9. West GB et al (2001) A general model for ontogenetic growth. Nature 413:628

3

Truth and Beauty

"Every day we slaughter our finest impulses. That is why we get a heartache when we read the lines written by the hand of a master and recognize them as our own, as the tender shoots which we stifled because we lacked the faith to believe in our own powers, our own criterion of truth and beauty. Every man, when he gets quiet, when he becomes desperately honest with himself, is capable of uttering profound truths." Miller [1].

3.1 A Journey to the Himalayas

Three things in life that most of us desire are love, money, and an abundance of good food and drink. However, the three things that we usually find are shattered illusions, obesity and death. Nevertheless, some people manage to transcend their peers and stand out for their higher aspirations.

Young Archibald was one of these few: his quest was for Beauty, Truth and Wisdom. To those who warned him that it was unrealistic to hope for all three, he would reply that actually the three were just different facets of the same thing. Beauty is perfection, and perfection cannot be flawed by the capital sin of being false. "Just think of a snowflake," he said. "The laws of physics and geometry render each flake different, but they are all endowed with the perfection of symmetry. What could be more enchanting than a snow-covered mountain landscape? Wisdom is the ability to recognize this union and be grateful for such a wonderful gift."

But where can one find beauty and truth? In the cities, the splendour of architecture, art, music and poetry, are offset by pollution, degradation and kitsch. Far better to search in isolated mountain valleys, in villages blessed by the stunning natural beauty of their surrounds, where simpler, but wiser communities prosper.

So Archibald travelled all summer and arrived in late autumn at a secluded village on the slopes of a beautiful mountain in the Himalayas. Since he was searching for wisdom, he was accepted by the Buddhist villagers, and even admitted to the weekly meeting of the ancients. In the first of these, the village Chief ordered that everybody should spend considerably more time collecting firewood for heating, for the predictions were for a very cold winter.

In the second and subsequent meetings, the Chief kept pressing for more and more wood to be collected, until one of the ancients protested and asked how the Chief could be sure that it would be so cold. The Chief did not elaborate how he knew, but said "if you don't believe me, visit the Wise Old Man in the mountain tomorrow and ask him."

The next day, Archibald and a delegation of ancients climbed for hours to reach the cave, high in the mountains, where the Wise Old Man was living alone. Archibald was elated: at last he would have a chance to meet the mystic, who in his eyes was the personification of wisdom. When they finally reached the cave, the old man waved them in and demanded their purpose.

As he heard the explanation, he nodded sagely. "Yes, your Chief is right," he said, "it will be a terribly cold winter."

But some of the ancients were still unconvinced: "How can you be so sure?".

At that the Wise Old Man replied gravely: "Because these past few weeks, down in the valley, I have seen many, many villagers relentlessly collecting firewood."

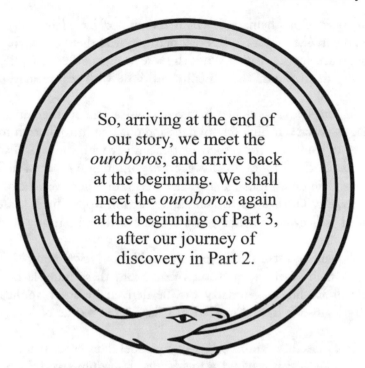

So, arriving at the end of our story, we meet the *ouroboros*, and arrive back at the beginning. We shall meet the *ouroboros* again at the beginning of Part 3, after our journey of discovery in Part 2.

The quest for truth and beauty is not confined to adventurers, such as Archibald but, perhaps surprisingly, is shared by scientists, artists, writers and lay people alike. However, not everyone means the same thing when they use these two terms. In this chapter we shall explore further what a scientist means by truth, and how the impact of concepts of beauty has influenced the direction of scientific research.

3.2   A Comparison Between Mathematical and Physical Truth

In Chap. 1, we explored the possibility of multiple truths, and observed that differences may arise between "scientific truth" and "theological truth." What is perhaps more surprising is that even within the sciences (if one includes mathematics in this category), clear differences of interpretation of what is meant by "truth" are evident. In this Section, we attempt to clarify some of these differences.

It is difficult to imagine any student progressing far in physics without a good grounding in mathematics. The logical nature of physics is closely related to that of mathematics, which undoubtedly explains why mathematics is so entwined with physics. Modern physical theories are formulated in abstract mathematics, which is not the case in most other disciplines. This gives physics its own aura that discourages those with scanty mathematical talent.

To make the correspondence between physics and mathematics a little clearer, let us revisit our high school days and our first introduction to geometry. For most students, this is their first encounter with the logical structure of mathematics. Euclid founded his geometry upon a number of definitions and axioms. The definitions tell us what we mean when we talk of points, lines, circles, etc. The axioms, which are detailed in Appendix 3.1, are sometimes called "self-evident truths", and are assumed without any attempt to prove them.

All Euclid's axioms seem to be very reasonable, and few among high school students, or their teachers, question them. From these humble beginnings the whole of Euclidean geometry can be derived by strict application of elementary logic. We have discussed in Chap. 2 the nature of logic, and its origins.

The fifth of Euclid's axioms states that parallel lines never meet. Over the centuries the question was raised as to whether this axiom was really necessary, or indeed, true. From a purely mathematical point of view, mathematicians attempted to deduce the axiom logically from the other four (in which case the fifth would have lost its status as an axiom, and become a theorem), but had no success. We live on a spherical planet, which, because of its size, appears to us to be totally flat locally (except for hills, etc.). Hence, we might expect from our experience that the fifth axiom could be valid locally, but not globally.

Let us, for instance, imagine two-dimensional animals living on a large sphere, analogous to the Earth. If two such individuals set off from different points on the equator on parallel trajectories heading due north, they will meet each other at the North Pole because of the curvature of their spherical domain. In their world, parallel lines do indeed meet. Eventually, nineteenth Century German mathematician, Bernhard Riemann, developed a geometry for such a world. As we shall see in Chap. 7, in Einstein's General Theory of Relativity, where space–time is not flat, Euclid's geometry does not apply and Riemannian geometry is essential.

To summarise, mathematics uses logic to deduce complex "truths" from underlying assumed simpler truths. In some branches of mathematics, these basic truths may not be all that intuitive. However, they represent the foundation, on which the towering edifice of any field of mathematics is constructed.

In the case of Euclidean geometry, the axioms reflect the reality of the terrestrial space in which we live our daily lives. However, it is not necessary for the axioms to have any connection with physical reality at all. Many fields of mathematics are highly abstract, and some mathematicians consider only mathematics that has no practical applications to be "pure". G. H. Hardy, whom we will meet again later in this Chapter, made a distinction between "real" mathematics, "*which has permanent aesthetic value*", and the remainder, "*the dull and elementary parts of mathematics*", that have practical use [2].

Hardy's aloofness provoked retorts, such as the following, from those of a more practical bent: *A group of physicists and engineers were enjoying a flight in a balloon until sudden gusts of strong wind compelled them to seek a safe landing place. After they had set down, a physicist in the party leaned out of the basket and asked a passing local where they were. "In a balloon", came the reply. The physicist turned back to his companions. "That man is obviously a mathematician," he declared. "His answer is totally correct, but absolutely useless."*

Let us now see how the axiomatic structure of mathematics compares with the nature of physics. In physics, we start with definitions of quantities, such as energy, momentum, velocity, mass, length, time, force etc., that are analogous to the definitions of Euclid. Some of these are more fundamental than others: for instance, velocity is defined as the length travelled in a particular direction divided by the travel time. Then, instead of axioms, we have physical "laws", which are inferred through observations and experiments. For example, Newtonian Mechanics, which describes the motion of objects in the everyday world, is based on three laws (see Appendix 3.2). From them and from the basic definitions, the interactions of bodies, both in the laboratory and in the heavens, can be calculated using the rules of logic, as embodied in mathematics.

Although the analogy between physics and mathematics is clear from the above arguments, there is an important difference. The axioms of mathematics are in a sense *a priori*, or originating from the mind of mathematicians.[1] There may be a correspondence between these axioms and the physical world, but it is by no means necessary. Likewise, Riemann developed his own geometry where parallel lines meet, not to describe travels on the Earth, but as a kind of mathematical game. When Einstein developed

---

[1] We have discussed in Chap. 2 the ongoing philosophical debate on the a priori versus a posteriori nature of mathematics (Platonism versus Empiricism). However, since practising mathematicians formulate their axioms with little or no regard for the physical reality, we presume axioms to be a priori for our purposes here. In Part 3 we shall reconsider this issue.

his Theory of Relativity, Riemannian geometry was already there, waiting for him to use as a framework for space–time.

The laws of physics, on the other hand, are a property of nature. They represent the physicist's attempt to understand the working of the cosmos. From these laws, the use of mathematics leads to predictions of the behaviour of interacting bodies that can be tested by experiments carried out in a laboratory, or by observations made through a telescope, or similar instrument. If these predictions do not agree with what is actually observed, it is assumed that the original laws are inaccurate and must be modified in some way, or abandoned altogether.

Strictly speaking we might alternatively say that the rules of logic that we followed are wrong, and some new logic must be devised. However, this alternative is seldom seriously considered. One might say that it is part of the "faith" of the physicist that the logic used by mathematicians is adhered to by nature. We will discuss this proposition further in Sect. 3.3 of this Chapter, and in Part 3 of this book we shall also examine the possibility of non-classical logics.

As soon as we mention experiments, we encounter the major difference between physics (or indeed any science) and mathematics. Every experiment, or observation, has associated with it a measurement error. For a well-designed and executed experiment, this error can be relatively small (say 0.1%), but it is always there. The consequence is that we can never verify our underlying physical laws once and for all time, because sometime in the future a more accurate experiment may reveal a small but significant difference between the predictions of our theory and the actual observations.

Thus we have finally arrived at the real difference between "truth" in mathematics and in physics. In mathematics, if a conclusion can be deduced as the logical outcome of a chain of reasoning traced back to the original axioms, it is considered true, and is announced to the world as a "theorem". In physics, there is no such thing as "truth": the best one can say is that within the bounds of current measurement error, the predicted result is correct.

3.3   Is Near Enough Good Enough?

The difference in attitudes between the two disciplines of mathematics and physics can be further illustrated by considering an arithmetical equivalence:

$$3987^{12} + 4365^{12} = 4472^{12}.$$

that came to popular attention when it was seen on a chalkboard belonging to Homer Simpson in the 1998 episode: "*The Wizard of Evergreen Terrace*" of the animated TV comedy, "*The Simpsons*". (The strange story behind Homer Simpson and his chalkboard is discussed in Appendix 3.3.)

Here, each of the three integers 3987, 4365 and 4472 is raised to the twelfth power.[2] If we try to check the accuracy of this relationship with our calculators, we will need to be inventive, as the individual terms exceed the largest integers ($2^{31}-1$) that can currently be stored in a normal computer. Let us now pose the question: is this equivalence correct, or not?

To a mathematician, the answer is straightforward: the relationship is wrong because it violates a theorem proposed in the margin of a book in 1637 by Pierre de Fermat, and now known as Fermat's Last Theorem. It took 358 years for a proof of it to be discovered by Andrew Wiles [3]. The theorem states that integer relationships, such as the one above, are not possible for powers greater than 2. A mathematician would not need to carry out any arithmetic to know that this relationship must be wrong.[3]

However, an actual numerical evaluation would reveal that the difference between the two sides of the equivalence is exceedingly small (about one part in a hundred billion), and quite negligible compared with the measurement errors that are observed in the most accurate experiments in a physics laboratory. To most physicists, the equivalence is therefore "true", and they would not hesitate to use it, if any of their theories required it. In this sense, near enough is good enough.

This attitude might seem reprehensible, but the aim of physics is to explain nature, as far as measurements allow it. Physical laws are constantly being refined as new experimental information comes to hand. Mind games are best left to mathematicians.

We have already encountered this different mindset in an example, which we discussed in Chap. 2. Newton's Law of Gravity cannot be solved exactly for a many-body system, such as our solar system, comprising the Sun, Earth, Moon, other planets and rocket ships. However, extremely accurate *approximations* to the solutions can be obtained, accurate enough to send space-crafts to distant planets. These approximations may not be *exact* solutions, but they are *near enough*.

---

[2] 3987^{12} is a shorthand way of writing $3987 \times 3987 \times 3987 \times 3987 \times 3987 \times 3987 \times 3987 \times 3987 \times 3987 \times 3987 \times 3987 \times 3987$.

[3] For those interested, a simple arithmetical disproof of the relationship is presented in Appendix 3.3, and the magnitude of the discrepancy between the two sides of the equivalence is estimated.

3.4   Faith in Physics

In a letter to Max Born, dated 29th April, 1924, Einstein wrote: "*I find the idea quite intolerable that an electron exposed to radiation should choose of its own free will, not only its moment to jump off, but also its direction. In that case, I would rather be a cobbler, or even an employee in a gaming house, than a physicist*" [4]. Einstein was expressing his opposition to the random, or probabilistic, nature of Quantum Mechanics. We shall discuss Quantum Mechanics in Chap. 5. However, what is relevant here is that Einstein was essentially expressing an article of faith, rather than a scientific argument. Today few physicists share Einstein's point of view. As we shall see, the experimental evidence in favour of Quantum Mechanic's weird predictions is just too all-encompassing to ignore.

If someone of the stature of Einstein could adopt such an "unscientific" attitude, which other articles of faith have become embedded in the science of physics without attracting overmuch critical comment? In the remainder of this Chapter we shall try to unearth a few, with the aim of encouraging our readers to seek out more for themselves.

Before beginning, let us recall for a moment a story from the field of mathematics. British mathematician, G. H. Hardy, was a child prodigy who could write numbers up to millions at the age of two years. As an adult, he spent most of his research life at Cambridge University, where he introduced an increased rigour into British mathematics, which at that time, just prior to World War 1, differed from European mathematics in the extent that it regarded rigour as relevant. By rigour we mean a strict adherence to formal logic, where literally nothing is taken for granted.

An example of rigour carried to the extreme is *Principia Mathematica* (PM) by Russell and Whitehead [5]. Their objective was to set mathematics on a sound formal basis. Their monumental work is hardly bedside reading, and we suspect is more often cited than read. However, one person who did read it, and was inspired by it, was Kurt Gödel, whose work we will discuss in Chap. 4. Another mathematician who realised its importance was G. H. Hardy.

As an example of the lengths to which Russell and Whitehead were prepared to go, we have included a small excerpt from PM in Appendix 3.4. After 362 pages of close mathematical reasoning, the authors have almost reached the stage where they can prove that $1 + 1 = 2$. As Whitehead himself remarked: "*It requires a very unusual mind to undertake the analysis of the obvious.*"

If Hardy was a mathematician who, like Whitehead and Russell, stressed the importance of formal proofs, Srinivasa Ramanujan was the exact opposite. Self-taught and a 23-year old shipping clerk from Madras (now Chennai), India, he wrote to Hardy in 1913, claiming that he had discovered some important results in Prime Number Theory, as well as other areas of mathematics. The ideas had come to him in dreams. Hardy and Ramanujan are depicted in Fig. 3.1.

Realizing that Ramanujan was no crank, Hardy invited him to Cambridge, and thus began one of the strangest collaborations in mathematics, which Hardy described as *"the one romantic incident in my life."* Ramanujan had little interest in the type of formal logic that Hardy regarded as essential. His ideas came to him in flashes of inspiration, or as he claimed, in visions from a Hindu goddess. It is unlikely that this assertion would have impressed Hardy, who was a confirmed atheist.

Fig. 3.1 Perhaps the strangest collaboration in the history of mathematics was between G. H. Hardy and Srinivasa Ramanujan. Hardy (left) was a child prodigy, atheist, and a product of the English public school and university system; Ramanujan (right) was a self-taught shipping clerk and a devout Hindu. Images are from Wikimedia Commons (Image 1 by Unknown (Mondadori Publishers) [Public domain], https://commons.wikimedia.org/wiki/File:Godfrey_Hardy_1890s.jpg (accessed 2020/8/31); Image 2 by Konrad Jacobs - Oberwolfach Photo Collection, original location, CC BY-SA 2.0 de, https://commons.wikimedia.org/w/index.php?curid=3911526 (accessed 2020/5/12))

Not all of Ramanujan's conjectures[4] turned out to be true, but many of them did, and they were ground-breaking in their originality. Hardy and Ramanujan worked together to develop formal proofs for them, an exercise that interested Hardy much more than Ramanujan, who preferred to approach his Goddess for further exciting titbits of information. This unusual collaboration was cut short by the untimely death of Ramanujan from consumption in 1920 at the age of 32. The story of this strange partnership has been narrated by Robert Kanigel [6]. Some of Ramanujan's highly unorthodox results have inspired much research by later generations of mathematicians, with applications in fields as diverse as String Theory and Crystallography. The reason for recounting this story here is to show how, even in the normally rigorous discipline of mathematics, different points of view can be adopted. Certainly no modern mathematician would feel comfortable with the *ad hoc* conjectural methodology of Ramanujan, and would not be satisfied until a formal proof of their conjectures had been established. However, how many are willing to go to the lengths of Whitehead and Russell in their search for rigour? This last question becomes even more relevant when we discover, in Chap. 4, that such a quest is doomed from the outset to failure.

3.5   Truth and Beauty

'Beauty is Truth, Truth Beauty,' - that is All Ye Know on Earth, and All Ye Need to Know.

The above lines from John Keats' "*Ode to a Grecian Urn*" have incited much debate in the world of English literature. Whether the poem can be classed as good or bad is not for us to discuss here. Suffice it to say that on that topic there is a multitude of different opinions.

A Sosibios Greek vase from the Louvre was one of the sources of Keats' inspiration. It is of a classic form that was regarded by the Greeks as possessing great beauty. Figure 3.2 below shows a terracotta amphora from the period ca. 525 – 500 BCE that demonstrates this classical shape. The cylindrical symmetry of the vase is elegantly broken by the two handles, which have their own bilateral symmetry about a vertical plane.

It may seem incongruous, in a book on physics, to talk about beauty, and yet, as we shall see in the following, beauty and symmetries have played their part in guiding scientific thought, and we shall see that analogies can be

---

[4]Conjectures in mathematics are statements which are believed to be true, but have not (yet) been proved.

Fig. 3.2 Terracotta Panathenaic prize amphora ca. 525–500 BCE, attributed to the Kleophrades Painter. On view Metropolitan Museum of Art, New York. Image: Public Domain (https://www.metmuseum.org/art/collection/search/247959 (accessed 2020/9/11))

drawn between the concepts of what is attractive in the worlds of art and science. In fact, this analogy has been carried so far as to suggest that physics and poetry are but two different paths in the discovery of reality [7].

The first, and possibly the most important, observation is that beauty to one person may well be ugliness to another. Helen of Troy was reputedly the most beautiful woman in the ancient world and her abduction led to the Trojan War. She has been depicted many times over the centuries, and these portraits provide an insight into what the artists (mostly men) at various times considered to be their ideal female form.

Figure 3.3 shows a portrait by Impressionist painter, Auguste Renoir, of a beautiful young woman of the Nineteenth Century. A visit to any major art

Fig. 3.3 *Young Woman Braiding Her Hair, 1876* by Auguste Renoir. Image: Open Access, courtesy of NGA, USA (https://www.nga.gov/collection/art-object-page.52207. html)

gallery will reveal that the classical nudes of Rubens (and later, Renoir) are much heavier than today's svelte fashion models.

Facial scarring of men and women is regarded as tragic in the western world. However, in some African tribes, ritual facial scarring is commonplace and considered beautiful in women and showing strong character and loyalty to tradition in men. The practice is now becoming less common as western values become all pervasive.

Music is another branch of art where tastes in beauty vary widely. Classical music may not be appreciated in a household brought up on popular genres, such as rock, country, and jazz. Even within the classical repertoire, there are disparities of style that do not appeal to all. The atonal music of Arnold Schoenberg leaves many repulsed by the strong discords, and it may take considerable exposure before one begins to appreciate the subtleties that lie beneath. Schoenberg is an "acquired taste", an attribute that he shares with many other "beautiful" things. Spicy food, the bitterness of beer and the dryness of wine are not tastes that the newly weaned infant is likely to enjoy. Rather, that appreciation is brought about by familiarity, born of experience.

What then characterises a beautiful piece of physics, and why should we care? After all, also for scientists *"beauty is in the eyes of the beholder"*, while science should be by definition objective. Sabine Hossenfelder in her provocative book, *"Lost in Math: How Beauty Leads Physics Astray"* has polled many modern physicists over what they consider a beautiful theory. She writes: *"When asked to judge the promise of a newly invented but untested theory, physicists draw upon the concepts of naturalness, simplicity or elegance, and beauty. These hidden rules are ubiquitous in the foundations of physics. They are invaluable. And in utter conflict with the scientific mandate of objectivity* [8]".

"Naturalness" implies that the free parameters in a theory are not fine-tuned to achieve agreement with experiments or observations. It also includes an aversion to theories which contain very large or very small values for the dimensionless constants (see Sect. 3.7). "Simplicity" will be discussed in the next Section. "Beauty" usually involves underlying symmetries in the mathematical equations. Symmetries allow the prediction of more results from fewer basic assumptions.

The pursuit of beauty is not just a recent interest in physics. The circle, being the most symmetric plane figure, requires only the specification of a central point and a radius to completely identify all of its properties. The early Greek astronomers took it as given that the motion of the planets, Sun and Moon were described by circles, which had for them almost divine properties. Ptolemy wrote: *"Our problem is to demonstrate in the case of the five planets, as in the case of the Sun and Moon, all their apparent irregularities as produced by means of regular and circular motions (for these are proper to the nature of divine things which are strangers to disparities and disorders).* [9]". We shall describe the Ptolemaic model of the solar system in more detail in Chap. 10.

This prejudice in favour of the circle persisted even after Kepler had shown that the planetary orbits were, in reality, ellipses. Kepler himself struggled to reconcile the observations made by Tycho Brahe with circular heliocentric orbits, before giving up and accepting that the orbits had to be elliptical. Galileo took no interest in Kepler's elliptical orbits: he maintained that *"only circular motion can naturally suit bodies which are integral parts of the Universe as constituted in the best arrangement"* [10].

Even in the twentieth century, beauty was an accepted criterion for the assessment of a physical theory. Endorsements of this criterion by famous scientists are easy to find:

"If nature were not beautiful, it would not be worth knowing, and life would not be worth living." – Henri Poincaré;

  "I have deep faith that the principles of the Universe will be beautiful and simple." – Albert Einstein;

  "A theory with mathematical beauty is more likely to be correct than an ugly one that fits some experimental data."– Paul Dirac;

  "My work always tried to unite the true with the beautiful; but when I had to choose one or the other, I usually chose the beautiful." – Hermann Weyl.

As we shall see in Part 2 of this book, the pursuit of beauty in physics has led to some spectacular successes. Dirac predicted the existence of a hitherto undiscovered particle, the positron,[5] just from the symmetry of his equations. Symmetries in the Standard Model of Fundamental Particles (which we shall discuss in Chap. 9), led to predictions of the existence of two new particles,[6] that were subsequently discovered.

As we have seen in the case of Ptolemy's planetary orbits, being beautiful does not necessarily mean that a theory is correct. Over the years since Kepler, astronomers have by necessity come to appreciate the appeal of ellipses, which once were regarded as not perfect enough to be part of the divine plan. Astronomical observations forced the theorists, in the eighteenth century, to give up their prejudices.

In the fields of Cosmology and Fundamental Particle Physics today, observations and experiments that can seriously test modern theories are difficult and very expensive. As a consequence, there is a lack of observational data to direct today's physicists away from the theories that they consider beautiful. Without the impetus given by observation and experiment, tomorrow's physicists may not be encouraged to overcome current prejudices and develop their own concepts of beauty. Progress in these fields may then slow. We shall discuss this issue further in Part 3 of the book.

3.6   A Monk with a Good Razor

The belief in the inherent simplicity of nature, if only we are smart enough to discover a few underlying physical laws, has been the driver behind physics for centuries. A genius in the world of art can capture much of the human condition with just a few lines drawn on a piece of paper, and we see an example of this in *Jan Cornelisz Sylvius, the Preacher* by Rembrandt, displayed

---

[5]A particle similar to the electron, but with positive charge. We will meet these in Chap. 8.
[6]The Omega Minus and the Higgs Boson.

in Fig. 3.4. Likewise, physicists hope to explain the physical world with only a few laws (or assumptions).

As assumptions may always be questioned, the fewer assumptions, the more confidence we can have in the conclusions drawn from them. This simple assertion is usually attributed to William of Ockham (c. 1287–c 1347), an English friar and theologian. However, the same, or a similar principle was already used by the ancient Greeks, e.g. Aristotle and Ptolemy. Many other thinkers followed, before and after Ockham, with similar assertions, but either luck, or the empathy of some later philosopher, has granted Ockham the honour of having his name associated with the criterion, which is generally known as Occam's (or Ockham's) Razor. It is a *rule of thumb*, rather than a scientific law. The basic idea is that from among competing hypotheses that explain an observed result, the one that requires fewest assumptions should be selected.

Fig. 3.4 *Jan Cornelisz Sylvius, the Preacher*, circa 1644–1645 by Rembrandt van Rijn, circa 1627–1628. Pen and brown ink on laid paper, National Gallery of America, Washington DC. Image: open access (https://www.nga.gov/collection/art-object-page.45998.html (accessed 2020/9/11))

In science, Occam's razor is used as a practical guide in the development of new theoretical models, but not as a criterion to establish their validity. In fact, it may happen that, later on, new observational or experimental data require us to modify or abandon these models in favour of new ones. It is important to remark that in science there is no place for dogmas, and that every new result must be seen only as one more step in a long journey, rather than as the final word.

Simplicity has of course obvious practical advantages, and also a powerful aesthetic appeal. The main criterion for the acceptability of a given theory is its *falsifiability*, i.e. the possibility to prove it wrong (see Appendix 3.5). This concept, introduced by Philosopher, Karl Popper, may be difficult to understand and even counterintuitive. However, if it is impossible to prove that a certain theory is wrong, we are left only with the alternative of believing it, or rejecting it altogether. Our decision becomes an act of faith, instead of being based on experimental or observational evidence, in open contradiction with the scientific methodology. Moreover, with fewer assumptions, the burden of falsifying them is lighter.

As an example, let us examine the Bishop of Ussher's[7] assertion that the earth was created at 6 p.m. on Saturday, October 22nd, 4004 BCE in the proleptic Julian calendar,[8] near the autumnal equinox. Which time zone this referred to is unknown. Presumably it wasn't daylight-saving time, as in Genesis the Sun wasn't created until the fourth day. When presented with evidence of fossils that can be dated to much earlier times than this, some fundamentalists claim that the fossils were also created, along with everything else, in 4004 BCE.

It is impossible to disprove in absolute terms such a thesis. If the fossils, along with all other prehistoric relics, cave paintings and the like, had been created in some kind of celestial scam in order to trick naïve scientists and archaeologists, then the scam cannot be falsified. Ussher's assertion, by its nature, can always be tailored to explain the observed data. However, according to Popper, such theories are not science.

It is, of course, easy to use Occam's Razor to counter the old bishop. Scientific theories are much simpler than Ussher's explanation, involving, as they do, only one Big Bang, compared with the requirement of countless individual acts of creation for each fossil and ancient artefact.

---

[7]James Ussher, Archbishop of Armagh (Church of Ireland), 1581–1656 AD.
[8]The proleptic Julian Calendar extends the Julian Calendar backwards in time, taking correct account of Leap years. The Julian Calendar is the precursor of the modern Gregorian calendar.

However, we come here to a point of contention. How do we count the number of assumptions in the alternative explanations of a scientific observation? Is an individual assumption required for every fossil and artefact, or will one overriding assumption for creation as a whole be sufficient? Do we require an individual assumption for every act of magic performed by a leprechaun (see Fig. 3.5), or does belief in the existence of leprechauns count as a single assumption? As we can see, the application of Occam's Razor is by its nature subjective.

We will in the course of this book come across many applications of Occam's Razor. It lies behind the fundamental desire to reduce physics to as few laws as possible. Perhaps the greatest achievements of 19th Century physics was the unification by James Clerk Maxwell of two separate phenomenologies, electricity and magnetism, into just one: electromagnetism. As a consequence of Maxwell's work, the number of physical laws required to explain a wide range of physical phenomena was significantly reduced. This success inspired physicists to attempt the unification of the

Fig. 3.5 Leprechaun, with his pot of gold. Does every act of magic he performs count as a single assumption when applying Occam's Razor, or does belief in the leprechaun count as only one assumption? Image from Wikimedia Commons (Image by Croker, T. C. (1862) Fairy Legends and Traditions of the South of Ireland. https://commons. wikimedia.org/wiki/File:Leprechaun_or_Clurichaun.png (accessed 2020/5/12))

other fundamental forces (weak nuclear, strong nuclear and gravitational) into a Theory of Everything (TOE). As we will see in later Chapters, they were only partially successful.

3.7   O Heaven, Were Man but Constant, He Were Perfect! [11]

Another consequence of Occam's Razor is the belief that the laws of physics and the fundamental constants they involve are unchanging throughout the breadth of the Universe, that they have remained unchanged throughout its history, and presumably will remain unchanged forever. A word of explanation is necessary here. The laws of physics often include a physical constant, the value of which must be obtained from measurement. An example is Newton's Law of Gravity, where the strength of the gravitational field is given by G, the so-called Gravitational Constant. Likewise, the speed of light in vacuum, denoted by c, appears in many formulas in any text book of physics.

Both G and c have physical dimensions, which means that their value depends on the system of physical units used to measure them. This is quite arbitrary. For instance, c is normally expressed in metres per second, but could just as well have been given in feet per second, or furlongs per fortnight. Experimentalists in nuclear physics often use the approximate value of one foot per nanosecond for c as a guide to help them lay out detection equipment in their time-of-flight experiments.

However, as we saw in Chap. 2, it is always possible to construct various products and ratios of physical quantities to produce dimensionless parameters. We can carry out this procedure with the fundamental physical constants. For instance, if we consider the ratio of the electron mass to the proton mass, we obtain a dimensionless number that is independent of any arbitrariness arising from the choice of units. This ratio will always have the same value, irrespective of whether we measure the electron and proton masses in kilograms, pounds, or any other units of mass. The value of this ratio has therefore a more fundamental physical meaning than the individual masses.

Another of these dimensionless fundamental constants is the Fine Structure Constant α which characterises the strength of the electromagnetic interaction between elementary charged particles. We shall discuss this constant further in Chap. 9. Normally its value, which is known with incredible accuracy (up to about 1 part in a billion) is expressed as a reciprocal, i.e.

$$1/\alpha = 137.035999037(91).$$

The Fine Structure Constant remains one of the great enigmas of modern physics. Nobel Laureate, Richard Feynman, stated that all theoretical physicists should go to sleep at night with its value pasted on the wall above their bed to remind them of its unexplained nature. He had this to say about the issue:

> "Immediately you would like to know where this number for a coupling comes from: is it related to π or perhaps to the base of natural logarithms? Nobody knows. It's one of the greatest damn mysteries of physics: a magic number that comes to us with no understanding by man. You might say the "hand of God" wrote that number, and we don't know how He pushed his pencil. We know what kind of a dance to do experimentally to measure this number very accurately, but we don't know what kind of dance to do on the computer to make this number come out, without putting it in secretly!" [12].

The very name of these quantities, i.e. *constants*, reflects the implicit belief that their values do not change throughout the past and future, or over the length and breadth, of the Universe. Likewise, the laws of physics are also believed to be immutable. The laws of Conservation of Momentum and Energy are found to be valid from sub-atomic to cosmological scales. This immutability of physical laws and constants is a credo for which Occam's Razor provides some sort of justification, i.e. that unchanging laws and constants provide a *simpler* scenario.

From time to time, prominent physicists, such as Sir Arthur Eddington and Paul Dirac, father of Relativistic Quantum Mechanics, whom we shall meet in Chap. 9, have floated the idea that some of the fundamental constants may change their values over time. The latter proposed that the value of the Gravitational Constant G decreases by 5 parts in 10^{11} per year as the Universe expands, a rate of change far too small to be detected by current terrestrial experiments. Philosopher and mathematician, Alfred North Whitehead wrote that the laws of physics themselves *must* change [13]:

> "Since the laws of nature depend on the individual characters of the things constituting nature, as the things change, then consequently the laws will change. Thus the modern evolutionary view of the physical Universe should conceive of the laws of nature as evolving concurrently with the things constituting the environment. Thus the conception of the Universe as evolving subject to fixed eternal laws should be abandoned."

The speed of light c is ubiquitous in physics; it appears in innumerable formulas in physics textbooks. Over the period 1928 to 1945, the measured speed of light fell steadily by an amount that was outside the bounds of quoted experimental errors. Then in the late nineteen-forties new measurements were made that were in better agreement with the 1928 measurements. Had there perhaps been a cyclic change in c over this time period? Most physicists believe not, and ascribe the result to a self-censorship process by experimenters. Measurements too far from the currently accepted value tend to be treated as statistical flukes and discarded. However, the possibility of a variable speed of light has been raised in alternative models of the early history of the Universe. We shall discuss this issue further in Chap. 11.

The long-term variability of other physical constants, e.g. Planck's Constant (see Chap. 5) and the electronic charge, has also been suggested. The question of whether the Fine Structure Constant α is varying now, or has changed in the past, is an active research field (see Appendix 3.6). Clearly these investigations have important implications for cosmological theories. However, it will take very convincing observational data to persuade physicists to abandon the assumption of constancy in the physical laws and their associated fundamental constants, partly because of Occam's Razor, but also because once the laws and constants are allowed to change, anything goes. It is very difficult to put observational constraints on what occurs in another part of the Universe or in the very distant past and future.

If Occam's Razor is invoked in some areas of physics in a search for simplicity, in others it appears to have been completely disregarded by the specialists. So we have the scenario of multiverses, where at every instant the future timeline of the Universe splits into an infinite number of possibilities. Another theory requires that space–time comprise eleven dimensions, while in Cosmology the presence of Dark Matter and Dark Energy would imply that 95% of the Universe is unknown. As we progress through this book, we will return to these proposals in more detail. One can only imagine what Einstein's reaction would have been to some of them.

References

1. Miller H (1964) On writing. New Directions Publishing Company, p 25
2. Hardy GH (1940) A mathematician's apology. Cambridge University Press, Cambridge
3. Singh S (2005) Fermat's last theorem. Harper Perennial
4. Born M, Einstein A, Born-Einstein Letters, 1916–1955, Letter No. 48

5. Whitehead AN, Russell B, Principia mathematica. Cambridge University Press, Cambridge, 3 vols, 1910, 1912, 1913)
6. Kanigel R (1991) The man who knew infinity. Washington Square Press
7. Evangelista LR (2020) Physics and poetry, two representations of reality (Colloquium: *Física e Poesia: Representações do Real?* Instituto de Física da Universidade de São Paulo)
8. Hossenfelder S (2018) Lost in math, how beauty leads physics astray (Chap. 1). Basic Books
9. Ptolemy: Almagest IX 2
10. Galileo G (1967) Dialogue concerning the two chief world systems, Ptolemaic and Copernican (reprinted: University of California Press, Berkeley, 2nd ed)
11. Shakespeare W, The two gentlemen of Verona: Act 5. Scene 4
12. Feynman RP (1985) QED: the strange theory of light and matter. Princeton University Press, p 129
13. Whitehead, AN (1933) Adventures of ideas. The Free Press

4

Gödel, Hawking and the Foundations of Physics

4.1 Alice in Gödel-Land

The little girl was exhausted due to the long journey she had completed from her home to Gödel-land. However, because of the excitement of being in such a new and strange place, she found she was unable to settle down.

"Mom", she said, "I can't sleep." To emphasise the point, she sat up, fluffed her pillow and tossed her head back down on the bed with a dramatic thump.

Her mother hid a smile; she knew where Alice was heading. "Do they have bedtime stories in Gödel-land?" Alice asked, as though the idea had just occurred to her.

"Of course, they do. Which one would you like?".

"Little Red Riding Hood." It was her favourite – for the moment.

"All right, but you really must try to sleep. We've got a busy day tomorrow." It was Alice's first visit to Gödel-land, and her mother had many treats planned for her. She settled back and her mother began the story. "Once upon a time, a little girl, whose name was Cinderella, lived with her two older step-sisters ...".

"No!" interrupted Alice. Her mother paused, and regarded her daughter, who was struggling to sit up again. "Mom, I don't want Cinderella, I want the story about Little Red Riding Hood!".

"Ah, Alice, but this is Gödel-land. Things are different here. We can't just restrict ourselves to the fairyland of Little Red Riding Hood. While we're here, we have to consider a wider fairyland, where Cinderella, Snow-white, Pinocchio and all those other characters live. Otherwise story-telling here isn't possible."

"But ... but ..." Alice was sure this could not be right. She had never heard of anything so silly, but her mother was very clever, and knew so much more than she did.

"Now let me go on, and do try to sleep." Alice settled back down to listen. "'I have some cookies here,' said Cindy's step-mother, 'and I would like you to deliver them to your Aunt, to her house deep in the woods.'".

"Mom, no!".

"Now what?".

"It was to her grandma, not her aunt!".

"I know, Alice, I know. But as I told you before, here in Gödel-land we can't just limit ourselves to the world of grandmas. We have to consider a wider world with all sorts of other relatives, and sometimes even friends. Now, if you don't let me continue – without interruption – we'll never finish." Alice pouted, but restrained herself with an effort, and her mother carried on. "So, Cindy set out through the woods, and halfway to her aunt's, what do you think she saw, in the middle of the path, barring her way, but a –".

"Wolf! A big, bad wolf." shouted Alice.

"No. Actually, it was a cat."

"A cat?".

"That's right. A Cheshire cat."

"It was a wolf. It's always been a wolf. You're spoiling the story. And it's Little Red Riding Hood, not Cinderella."

"Now, Alice," said her mother, beginning to lose her patience, "You're not listening. In Gödel-land, we can't limit ourselves to the world of wolves –".

"I know, I know. I suppose we have to consider the wider world of all animals. But I don't like it. I don't like it one little bit! It's so confusing. Why do we have to do silly things like that? Why?".

"Do you really want to know?".

"Yes!".

"All right then, I'll try to explain."

Her mother pulled her armchair closer to Alice's bed, rearranged the cushions, and settled to begin what she knew would be a very long and complicated narrative. After just a few words, however, when she next glanced down at her daughter, she noticed that the little girl's eyes were closed. Alice was fast asleep.

4.2   The Price of Cakes in Gödel-Land

Our next goal is to try to explain what Alice's mother could not, i.e. the odd customs and rules of Gödel-land. However, first we need to abandon the land of fairy stories, and enter into an altogether different abstract dominion, the mathematical world of numbers. To begin, let us clarify what we mean by a number. The reader may think this is unnecessary. Surely, a number is just one of those quantities, like time and temperature, that everybody understands, but nobody seems quite able to define clearly (try asking around).

Nevertheless, the concept is not really so difficult: if you have a collection (or group) of any objects whatsoever and you disregard the nature of the objects, the number is the only characteristic that remains of the group. For instance, if you collect 2 pebbles, 3 cookies and one mango, and if you do not care what the objects are (i.e., pebbles, cookies or mangoes), the only thing that describes your collection is the number 6 (6 objects) …but don't eat the pebbles! The numbers, thus defined, are called integers (or natural numbers), and are contained in one small portion of what we may call the *number world*.

As trade began to flourish between humans, it became clear that the world of integers was insufficient to encompass transactions when debt was incurred by some individuals. In this case instead of owning 3 cookies in their collection, the debtors might owe 3 cookies to their neighbour's collection, with the two parties agreeing that the debt would be paid off some time in the future. The number world was therefore extended by including negative numbers to describe the debt. Much later, the number zero was included in the number world.[1]

---

[1]It might be interesting to recall that the zero was "invented" about 600 CE by the Indian mathematician, Brahmagupta, but it was introduced to the Western world by Leonardo Fibonacci only in 1228 CE (in his *Liber abbaci*).

Problems of the type: *what is the price of one cookie, if it costs $12 to buy four?* are common in commerce. The answer is, of course, $3. Suppose now that we replace "cookie" in the above question with some other object, e.g. liquorice stick. The answer is still the same, i.e. $3, irrespective of what object we choose. We can show this independence of the nature of the object by writing the question in the form: *what is the price of x, if it costs $12 to buy 4x?*, or even more succinctly by: *Find x if 4x = 12.* The answer is $x = 3$ (i.e. 12/4) dollars, if our unit of currency happens to be the dollar. What we have introduced above is the concept of an equation (where two quantities, one of which is unknown, are located on opposing sides of an equals sign. The rules for solving the equation (i.e. finding the value of the unknown x) are contained in elementary *algebra*.

So far, so good, and for the moment our integer number world seems adequate for our purposes. However, what happens if four cookies cost $10? (i.e. $4x = 10$). We easily see that there is no integer in our number world to tell us how much one cookie costs. We do not need a calculator to find that x = 10/4 = $2.50. But 2.50 is no longer an integer. In other words, the number world of integers is too limited to include the solution of our new equation. To solve it we must go into a much wider number world, the world of "*rational*" numbers, which includes integers and fractions, such as 10/4, or 2.50.

Let us now assume that our problem is even more complicated, and we are asked to find a number (again let us call it x) such that when it is multiplied by itself (or *squared*) we obtain the value 4. We can write this in the form: $x^*x = 4$, or more concisely: $x^2 = 4$. Solving this equation presents no great problem, and we see that $x = 2$, because the square of 2 is 4, or alternatively, the *square root* of 4 is 2. (Another possible solution to this equation is x = − 2).

Now, suppose we ask ourselves what is the square root of 2? This is a perfectly legitimate question. Imagine that we construct a triangle having two sides of unit length with a right angle (i.e. an angle of 90°) between them, then the length of the third side (i.e. the hypotenuse) is equal to the square root of 2, which turns out to be 1.414213562373095... (See Fig. 4.1.) Here we have arbitrarily truncated the number of decimal places to 15. In fact, there is no limit to the number of decimal places, and we could have filled the rest of the book with them. If we had, we would not have found any repeated

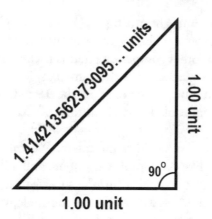

Fig. 4.1 A right-angled triangle: the length of its hypotenuse is expressed as an *irrational number*

patterns in the order of the digits. Such a number cannot be expressed as a fraction,[2] and is called an *irrational number*.

In other words, to find the solution of the equation $x^2 = 2$, one must go into a still wider number world, the one of "*real*" numbers, which includes both the *rational* and the *irrational* numbers. By the way, the word "irrational" conveys a deep apprehension of these numbers, which were not wanted by their discoverers (the school of Pythagoras) because they contradicted their faith in a numerological "purity". According to legend, the discoverer, a student of Pythagoras, was forced to commit suicide, lest his sacrilegious discovery be propagated to others.

Surely, now at last, our number world is sufficiently broad to allow us to carry out all the numerical operations we are ever likely to encounter. Not quite. For instance, take up your calculator and ask it to calculate the square root of −1 (or any other negative number). You will obtain a fairly dismissive error message from your machine, as if you had done something rather stupid. However, negative numbers are a legitimate part of our number world, and taking the square root is a legitimate mathematical operation. Surely one might expect a valid result from such an endeavour.

Mathematicians certainly thought so, and introduced into the number world a quantity i which they defined to be the square root of −1 (so that $i * i$, or $i^2 = -1$). The square root of −4 then became $2i$. Numbers of this type were deemed *imaginary*, because mathematicians thought that i was only a "trick" to solve the equations. The number world would now seem to have

---

[2]The converse is not true; some fractions can indeed be expressed as an infinite number of decimal places, e.g. $1/3 = 0.333333...$ and $1/7 = 0.\mathbf{142857}142857\mathbf{142857}...$, but in this case there is a pattern of repetition in the string of digits.

two separate divisions, one comprising real numbers and the other imaginary numbers. However, there are also *complex* numbers, which have one component that is real and another component that is imaginary. Complex numbers were for many years regarded with the same suspicion that Pythagoras had accorded irrational numbers. However, in the 19th Century it was realized that they were essential tools for the solution of many problems in science and engineering.

So, at last, our number world is complete. We can perform any mathematical operation we like on complex numbers and still find a result that is not outside the domain of complex numbers.[3] If you replace "telling a story" with "solving an equation", you can now see the analogy between Alice's story and the contents of the current section: both cases illustrate the need of going to a wider system. In the next section, we'll see what all this has to do with Gödel, and with some of the current developments of Physics.

4.3   Gödel and Completeness

Kurt Gödel (see Fig. 4.2) is considered by many mathematicians to be the greatest logician since Aristotle. Einstein had once famously said that his only reason for going to Princeton was the opportunity to walk there with Gödel and discuss problems with him. To illustrate his fame, it may be appropriate to recall Einstein's comments, when he handed out the first Einstein Award, assigned jointly in 1951 to Julian Schwinger and to Gödel. To the former he said "You deserve it" and to the latter "You don't need it", meaning that Gödel's scientific relevance was superior to the importance of the prize itself [1].

It is curious that something similar had been said thirty years earlier about Einstein himself, when the Nobel Prize judges hesitated before assigning the prize to him, since their policy had always been to reward only experimental discoveries or experimentally verified theories, which was not yet the case with Einstein's General Relativity. Eventually a way around the impasse was found by giving the Nobel Prize to Einstein for his work on the photoelectric effect. On that occasion one of the judges remarked that assigning the prize to him was an honour more for the prize than for Einstein himself. However, although Einstein is still something of a popular cult figure more than 60 years after his death, Gödel's name and work are largely unknown, even among many scientists and engineers.

---

[3]The reader might be tempted to suppose that determining the square root of i would require a further extension of our number world. Not so. The square root of i is $(1 + i)/\sqrt{2}$ (i.e., a complex number), as can be verified by squaring this value, which yields i once more.

Fig. 4.2 Kurt Gödel, one of the most significant logicians of the twentieth century, as a student in Vienna in 1925. Image: Wikimedia Commons (Public domain - Kurt Gödel Papers, the Shelby White and Leon Levy Archives Center, Institute for Advanced Study, Princeton, NJ. https://commons.wikimedia.org/wiki/File:Young_Kurt_G%C3%B6del_as_a_student_in_1925.jpg)

Kurt Gödel was born in 1906 in Brno, Slovak Republic, which at that time belonged to the Austro-Hungarian Empire. He studied in Vienna, where in 1929 he completed his Ph.D. in mathematical logic. His dissertation led him to his *completeness* and, later, *incompleteness* theorems, which he published two years later and which represent his most relevant scientific achievements. In 1933 he became lecturer at the University of Vienna, where he remained up to 1940, when he lost his position because he had many Jewish friends, and moved to the USA, where he became Professor at the Institute of Advanced Studies (IAS) in Princeton in 1953. He remained in Princeton until 1976, two years before his death (1978).

Perhaps, from a human point of view, the cause of his death is almost as puzzling as his beloved paradoxes. Late in his life, he developed severe paranoia, and would eat only food prepared for him by his wife, Adele, for fear of being poisoned. When she became hospitalised for six months in 1977, he refused to eat, starving himself to death.

Let us leave to psychologists the task of fathoming Gödel's complex personality. His introvert nature was doubtless much tested by the murder of a friend during the Nazi period and by the upheaval, both enthusiastic and acrimonious, that followed his discoveries. Instead, let us see if we can obtain an inkling of the nature of the work that so impressed Einstein (and others), and the implications that still resound through mathematics and physics.

As we have discussed in Chap. 2, the aim of formal mathematics was (and still largely is) to begin with a small number of axioms and definitions, and by using the rules of logic, to deduce ever more complex truths, or theorems. Probably the most extreme example of this approach is the *Principia Mathematica* of Russell and Whitehead, which we cited in the previous Chapter.

One of the banes that can occur in logic is the discovery of a paradox, which may indicate either that we have made a mistake in our reasoning, or that our logical framework is insufficient for our purposes, as we saw in the last Section when attempting to find the square root of 2 while limiting ourselves to rational numbers.

In Chap. 1, we encountered Zeno's Paradox of Achilles and the Tortoise. Another famous example, according to tradition, is due to Epimenides of Crete, who came up with the statement "All Cretans are liars".[4] Now this statement presents an intrinsic self-contradiction, since, if it is true, it cannot be asserted by a Cretan, because a Cretan would lie, and vice versa if it is false. Since Cretans, like everybody else, are not necessarily always lying or always truthful, the so-called "Liar paradox" is usually stated more succinctly as "This statement is false". If the statement is true, its content indicates it must be false, and vice versa. Now, since language is essentially a tool of communication, rather than of logic, it is not surprising that it contains contradictions or inconsistencies. What is really surprising is that the same thing also happens in mathematics.

While working on *Principia Mathematica*, Bertrand Russell uncovered a paradox in Set Theory, which he was using as the formal mathematical framework for his ambitious opus. Normally first encountered by mathematics students in advanced University courses, Set Theory enjoyed a brief period of popularity in the 1960s when it formed the basis of "*New Math*" and was taught to primary school students to help them learn arithmetic.

---

[4] All this, of course, has nothing to do with the truthfulness of Cretans, since it is only used as a dialectic argument. It is curious that St. Paul took the statement literally and used it, together with other arguments, to conclude that non-Christian Cretans were "evil beasts".

As satirist/mathematician, Tom Lehrer, stressed: "*the important thing is to understand what you're doing, rather than to get the right answer.*" New Math has long been confined to the dustbin of history as another failed experiment by misguided educationalists. However, Set Theory is still an important component of the foundations of mathematics.

Russell's Paradox (see Appendix 4.1) is of a self-referential nature, similar to the Liar's paradox above. Obviously it is disconcerting to base a formal analysis of the structure of mathematics on a theory with a paradox at its heart. Various solutions have been proposed for the paradox. However, what is of interest to us here is that it inspired a young student by the name of Kurt Gödel to investigate the logical foundations of mathematics and the nature of mathematical proofs. His discoveries have shown that the formal derivation of all of mathematics from a few basic axioms, as attempted by Whitehead and Russell, is actually impossible. Let us try to be a little more specific, without going into technical details.

Two things we require from any branch of mathematics are *consistency* and *completeness*. In the case of Whitehead and Russell, it was arithmetic that they attempted to put on a rigorous formal basis. By consistency, we mean that *it should not be possible*, by following two different chains of reasoning from the same axioms, to prove that something is both true and false. By completeness, we mean that if something is true, it should be possible to prove that it is true, starting from our axioms. There should not be any theorems that cannot be proved. We may not be smart enough to prove them at the moment, but we should have the hope, sometime in the future, of discovering a proof. It should certainly *not be the case* that they can *never* be proved.

However, what Gödel succeeded in proving is that, if we have a system that is consistent (i.e. does not contain any contradictions), then it must contain at least one statement that is true, but which can never be proved. (See Appendix 4.2).

Not a problem, we might say. Let us just include that unprovable statement as an extra axiom, and all will be well. Not so. The inclusion of the extra axiom will change the system so that somewhere else, another different, unprovable but true, statement will be discovered. It is like trying to plug the leaks in a rusty rainwater tank. The stresses of the repair operation usually cause several new leaks to appear somewhere else in the tank (see Fig. 4.3). Gödel showed that arithmetic (and presumably other branches of mathematics) cannot be both complete and consistent simultaneously.

We see from the above why Gödel's work has severe implications for mathematics. Indeed, debate still rages in philosophical circles as to whether it demonstrates that there are "ineluctable limits to human reason" [2]. Leaving

Fig. 4.3 Trying to add an axiom to a branch of mathematics to "prove" an unprovable theorem is like trying to repair a leaky rainwater tank

such lofty considerations to one side, in the next Section we shall explore the relevance of Gödel's work to physics.

4.4   Hawking and His Epiphany

If there is incompleteness in mathematics, can we also expect to find an analogous kind of incompleteness in physics? The hope of many eminent physicists at the end of the twentieth century, following a flurry of far-reaching advances in the fields of Quantum Mechanics, General Relativity and High Energy Physics (amongst others), was to discover an all-encompassing theory that would unite these disparate fields. They called such a theory a TOE, or Theory of Everything. In analogy with what Gödel has shown to be the case in mathematics, are we now to suspect that there may be contradictions in physics that can never be resolved, or observed phenomena that we can never explain within the laws of physics? In other words, are there fundamental limits to our knowledge, i.e., some *prohibition* that prevents us from ever arriving at a TOE?

Certainly there are areas of physics, and we will encounter some of these in Part 2 of this book, where there are very definite limits to our knowledge. One example is a *black hole*. We shall discuss these strange entities in more detail in Chap. 7. Suffice it to say here that they are extremely dense objects,

Fig. 4.4 Stephen Hawking (1942–2018). (Image public domain, courtesy of NASA StarChild Learning Center https://commons.wikimedia.org/wiki/File:Stephen_Hawking. StarChild.jpg (accessed 2020/9/1))

with gravitational fields so strong that *nothing* (not even light) can escape from their clutches, if it chances to venture too close. The central region of a black hole thus seems always destined to remain out of reach of scientific observation, and therefore beyond the limits of physics.

However, even if black holes represent an instance of incompleteness of our knowledge of the universe, they were certainly *not* predicted, or discovered as a consequence of Gödel's theorem. (For that matter, neither were the irrational and complex numbers, which appeared, as uninvited guests, long before Gödel's time). The eventual consequences of Gödel's theorem in physics stem from the realization that physical theories are essentially mathematical models. Hence, if some mathematical results cannot be proved, then surely there must also be some physical theories that cannot be proved.

In fact, in a famous lecture [3], Stephen Hawking [4] (see Fig. 4.4), one of the most brilliant theoretical physicists after Einstein, recounts his *conversion* from being a convinced advocate of TOEs to a sceptic, more inclined towards accepting that a version of Gödel's theorem also applies in physics. In his own words:

> "*Up to now, most people have implicitly assumed that there is an ultimate theory that we will eventually discover. Indeed, I myself have suggested we might find it quite soon. However, M-theory[5] has made me wonder if this is true. Maybe it is not*

---

[5]M-theory, a type of string theory, and a candidate suggested as the basis of a TOE, is discussed in Chap. 9.

possible to formulate the theory of the universe in a finite number of statements. This is very reminiscent of Gödel's theorem. This says that any finite system of axioms is not sufficient to prove every result in mathematics."

The concept is reiterated at the conclusion of his lecture:

"Some people will be very disappointed if there is not an ultimate theory that can be formulated as a finite number of principles. I used to belong to that camp, but I have changed my mind. I'm now glad that our search for understanding will never come to an end, and that we will always have the challenge of new discovery. Without it, we would stagnate. Gödel's theorem ensured there would always be a job for mathematicians. I think M-theory will do the same for physicists."

In other words, Hawking concludes, on a note of optimism for the new generations of young physicists, that a TOE would be bad news for physicists looking for a job: much better if there is no TOE and, as a consequence, always something new to discover.

Hawking was not the first to question the impact of Gödel's theorem on physics. Stanley Jaki in his book *"The Relevance of Physics"* [5] states that a TOE must be a consistent non-trivial mathematical theory, and hence, following Gödel, incomplete. Freeman Dyson maintains: *the laws of physics are a finite set of rules, and include the rules for doing mathematics, so that Gödel's theorem applies to them."* [6] On the other hand, many physicists find the use of Gödel's theorem in this context to be unconvincing, and that in an infinite universe, there will always be some things that cannot be proved.

If Hawking's conjecture is correct, how do we begin the search for evidence of Gödel's incompleteness in physics? We cannot simply seek out some area that is at the moment poorly understood, since it is more likely that our ignorance depends only on the fact that we have not yet discovered the right theory. This has always been the case in the past, before any of the great discoveries of science.

More interesting is when we are faced with a theory, such as Quantum Mechanics (see Chap. 5), that seems to work extremely well, since it fits all the observations or experimental data, and yields accurate predictions, but which defies our common sense. Here the question arises: is this evidence of some incompleteness of the theory, or is it only a consequence of the fact that our common sense has evolved in an environment markedly different from the realm of Quantum Mechanics? Indeed, in our everyday life we are in contact only with macroscopic objects and phenomena, while Quantum Mechanics helps us deal with a quite different microscopic world.

In the previous chapters we have learned that both our senses and our reason can be very deceitful, and that our deeply-held intuitions of concepts such as truth and reality may be dubious, or even fallacious.

Hence, if there exists indeed some incompleteness in the sense of Gödel, where should we turn to look for it? In 2015, three mathematicians and computer scientists, applying an admittedly simplified model, found that the problem of determining whether there is an energy gap between the lowest energy levels in a material is "undecidable" [7]. We shall explore what we mean by energy levels in Chap. 8. However, for our purposes here, suffice it to say that this issue is of major importance in the understanding of the physics of superconductors. When we say that the question is undecidable, we mean that it is impossible to find an answer, no matter how hard we try. This is a direct consequence of Gödel's theorem.

So, is there any chance that Gödel's theorem might account for some of the great puzzles of modern physics, that we are presently compelled to sweep under the carpet, such as the inconsistency between Quantum Mechanics and Relativity? We have now reached the point where we can proceed no further towards the answers to these questions without a deeper understanding of modern physics, and the limits to knowledge that it has revealed.

In Part 2 of this book, we make a brief visit to the various coalfaces of current physics research, where thousands of dedicated workers are chipping away in search of the occasional diamond. However, it is not our goal to take part in this venture ourselves here, since we have insufficient time and resources available for that purpose. Instead, our role will be to guide the interested spectator.

To enjoy the beauty of a Beethoven sonata, one does not require a degree in the theory of music, even if such a qualification may add a deeper insight to our appreciation. Nor are all the audience members, on their feet at Stratford-on-Avon after a stirring performance of Hamlet, graduates from a Stanislavsky school of acting. Similarly, it should be possible for non-specialists to obtain an appreciation of the important problems being addressed by physicists today, and have a spot of fun for themselves along the way. This is what we shall attempt in Part 2. In Part 3 we shall return to tackle once more the questions raised above, taking advantage of what we will learn in Part 2.

References

1. Wuppuluri S, Ghirardi G (eds) (2016) Space, time and the limits of human understanding. Springer, Berlin
2. Nagel E, Newman JR (2001) Gödel's proof. New York University Press
3. Hawking S (2002) Gödel and the End of the Universe (lecture). https://www.hawking.org.uk/godel-and-the-end-of-physics.html. Accessed May 13 2020
4. Hawking S (2013) My brief history. Bantam Books
5. Jaki S (1966) The relevance of physics. University of Chicago Press
6. Dyson F (2004) New York review of books, May 13, 2004
7. Castelvecchi D (2015) Paradox at the heart of mathematics makes physics problem unanswerable (Nature News, 09 December, 2015). https://www.nature.com/news/paradox-at-the-heart-of-mathematics-makes-physics-problem-unanswerable-1.18983. Accessed 13 May 2020

Part II

Today's Frontiers

5

The Incredible Quantum Mechanics

5.1 The Strange Tale of the King of Neutrinia

Some physicists are much taken with parallel universes and are happy to postulate their existence, expound on properties they might have, and rejoice in the fact that although there is little or no possibility of verifying the existence of these worlds, at least there is no inconsistency with physical theories that actually precludes their being.

In one of these multitudinous parallel worlds, there existed a tiny kingdom, called "Neutrinia", with a king, Neutrino the First, who was very young and inexperienced, with many in his kingdom claiming that he lacked sufficient gravitas for his regal role. Now King Neutrino was madly in love with Princess Juliet, the daughter of the king of Protonia, a neighbouring country. Juliet also loved Neutrino dearly, and the only requirement for their lasting happiness was the blessing of Juliet's father, the curmudgeonly King Proton the 100th.

King Proton was very proud of his long dynasty and of his kingdom's wealth. When King Neutrino asked for Juliet's hand, he replied that he would only concede his daughter to a King able to donate to her as a Wedding present the famous pink diamond, Darya-ye Noor. Unfortunately, Neutrinia was small and even if Neutrino commandeered its entire wealth, it would not suffice to purchase such an illustrious jewel. The distraught king approached one of his friends, Paolo Diracus, who happened to be a very famous physicist, for advice.

The answer he received was indeed very helpful. Diracus replied: "As you probably know, in this world, Heisenberg's Uncertainty Principle is valid, not just in the realm of physics, but also in the field of economics." King Neutrino nodded sagely, although he understood little. Technical issues tended to put him in a spin. The physicist continued: "And my experimentalist friends tell me that lately some of the fundamental physical constants have begun to fluctuate wildly in value. Indeed, h, a constant named after my colleague, Maximus Planck, has recently increased greatly. So perhaps you can borrow, free of charge, the amount of money you need to buy the diamond, provided that you only need to show it to Juliet's father for a very short time." Pulling a pencil and an old bus ticket out of his pocket, Diracus exclaimed: "let's do the calculations" and began to scribble down some equations and numbers.

To King Neutrino's delight, everything worked out exactly as Diracus had foretold. Neutrino held possession of the diamond just long enough to show it to Proton, and Neutrino and the beautiful Juliet were married and lived happily ever after.

* * *

If King Neutrino's world seems strange to you, dear reader, then welcome to the crazy domain of Quantum Mechanics. In this chapter, ideas that seem bizarre based on our everyday experience are explored, and shown to be not only possible, but essential for the explanation of experimental observations. "Common Sense", which was never common[1] and seldom sense, is no longer useful as a guide to understanding Quantum Mechanics. Instead reliance must be placed on consistency, meticulous experimentation and observation to lead us in our quest.

5.2   *Et Lux Fuit* (and There Was Light)

Let us continue our story in our own universe at the end of the Nineteenth Century, when the general consensus among scientists was that all relevant

---

[1]This observation is usually attributed to Voltaire: See Appendix 5.1. Suffice it so say here that common sense, in the domain of Quantum Mechanics, is of very little use, as we will discover in the following pages.

Physics had by now been discovered, with only a few details remaining to be clarified. The ground for such optimism was the recently proposed theory of electromagnetic fields and waves by James Clerk Maxwell. In his seminal papers and book [1], Maxwell provided a convincing explanation of all known electric and magnetic phenomena. His theory, which also gathered up earlier work by Faraday and Gauss, predicted many as yet unknown effects,[2] which were subsequently discovered. In addition, Maxwell succeeded in unifying two forces (electric and magnetic), which up to then had always been considered distinct and independent. Maxwell's achievement has been called *"the second grand unification"*, the first one being Newton's unification of terrestrial and celestial mechanics.

But perhaps the greatest merit of Maxwell's work was to show that light (or, more precisely, the propagation of optical waves) was nothing but the solution of his equations in the most elementary case of a completely empty medium (vacuum). This was remarkable, since all other cases of propagating waves require a medium of some substance to support them, e.g. air for acoustic waves (sound) and water for sea waves. Luckily, light can also propagate, although at the cost of some absorption, in tenuous media such as air. This applies not only to visible light (i.e. radiation in the range of frequencies which can be captured by the human eye). It also applies to x-rays, gamma rays, ultraviolet and infrared light, as well as to waves used for the propagation of radio and television channels. All these forms of radiation propagate in exactly the same manner, as predicted by Maxwell's equations, and are distinguished only by their different frequencies.

Frequency, along with *wavelength* and *velocity*, are the three quantities that characterise all forms of wave motion. When we throw a pebble into a pond, we can observe these properties readily. By concentrating our attention on one point in the pond, we can see the water surface rising and falling as the wave moves by. The number of times the surface rises and falls per second at this point is the frequency of the wave. On the other hand, if we concentrate on the wavefront itself, and measure how far the wave progresses in one second, we have the velocity of the wave. The third quantity, the wavelength, is the distance between successive wavefronts. A little thought will convince the reader that these three quantities are not independent; in fact, the velocity is equal to the product of the other two.

For some types of waves, the velocity is different for different frequencies. An example is ocean swell. These long wavelength waves are generated by

---

[2]For example, in 1887 Heinrich Hertz discovered microwave electromagnetic radiation, as predicted earlier by Maxwell.

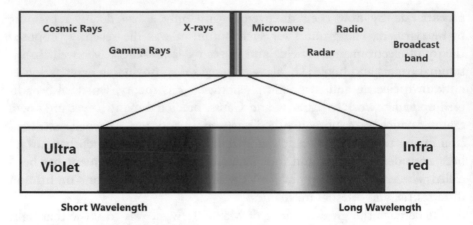

Fig. 5.1 The visible region, shown in colour, is only a small part of the overall electromagnetic spectrum, which ranges from cosmic rays to long-wavelength radio waves

distant storms, far from the beach where they are observed, and can sometimes travel half-way around the globe. The longest wavelength waves travel faster and arrive first, followed by those with progressively shorter wavelengths. The shortest wavelength waves, or *chop*, are soon absorbed and do not travel over inter-continental distances. A measurement of the wavelength gives an indication of how far away the storm that generated the waves was located. This characteristic—the variation of the velocity of waves with frequency—is called *dispersion*.

Electromagnetic waves travelling in a vacuum are non-dispersive, i.e. all frequencies travel with the same speed. This speed, called the speed of light, is one of the fundamental physical constants and is given the symbol c. However, when travelling through a medium, the waves are slowed down, and the magnitude of this effect does depend on the frequency.

In one of the most significant instances of *universality*[3] in Physics, the range (or *spectrum)* of naturally occurring electromagnetic frequencies spans as much as 14 orders of magnitude, i.e. 10 to the power 14, with the portion detectable by our eyes being almost inconsequential. The full electromagnetic spectrum is displayed in Fig. 5.1. Of course, in such a huge interval, there is an equally large variation in the physical properties and applicability (also

---

[3]When studying different fields of physics, it often happens that we find very similar patterns in the observed phenomenology, i.e. similar effects in totally unrelated situations. This "universality" is due to the fact that the basic equations are the same, even if they describe very different physical phenomena.

potential danger), but still the nature of electromagnetic waves, as predicted by Maxwell's equations, remains the same.

From the above discussion, it is clear that light is a form of wave propagation. No room is left for any doubt. As if these theoretical considerations are not convincing enough, there exists a rich phenomenology to demonstrate this particular characteristic: diffraction, interference, and polarization are all examples of light's *wavy personality*. These effects are discussed in more detail later in this Chapter. They can produce flashy displays of natural splendour, such as the rainbows that accompany a sun shower.

In Fig. 5.2, the superimposed reflection and refraction of light at the surface of a soap bubble produces an image of startling beauty. The reflection of the landscape is enhanced by the rainbow colours which result from refraction of the white light at the bubble surface. The refraction process disperses the white light into its constituent colours because light of different frequencies travels at different speeds inside the thin film comprising the bubble.

Despite the evidence outlined above for the wave nature of light, we shall see in the next section that there are other observations, from which it is

Fig. 5.2 A soap bubble, displaying the superimposed effects of reflection and refraction. Image courtesy of Jooinn (https://jooinn.com/colorful-soap-bubbles.html (accessed 2020/5/20))

equally clear that light is composed of beams of particles. Such is the weird and almost contradictory nature of Quantum Mechanics.

5.3   One, None, and a Hundred Thousand

In one of his best-known novels [2], *Uno, Nessuno, Centomila*, the Italian writer Luigi Pirandello argued that, although we may believe in our own enduring identity, we are in a constant state of change, not only because of the unavoidable effects of ageing, but also because we are different when we are with different people. We might change consciously: in fact, it may be advantageous for a young person to show off in the presence of a desirable potential partner, and then try to look miserable when asking for financial support from parents. But we also change unconsciously, for when we are in different company, we instinctively tend to distort our personality.

This behaviour is, of course, only human. Or is it? If we look at the nature of light, we might get the impression that light is no less *Pirandellian* than we are. In fact, one of the few details left to be clarified at the end of the 19th Century was precisely that: the nature of light. As we have seen in the previous Section, the wave-like character of light was very well established by Maxwell's theory.

On the other hand, since antiquity poets and scientists alike have spoken of solar and lunar rays. Newton's corpuscular theory of light asserts that light is made up of small discrete particles (corpuscles), which travel in straight lines (rays). The idea of tiny elementary particles comes from the first Greek philosophers who, even though fascinated by the immensity of the heavens, also turned their attentions to the very small.

Objects around us are usually complex, being constructed from various materials that are obviously different from each other and from the composite that they form. However, suppose we isolate one of these constituents, for example, a cube of sugar. What happens if we break it into two? Will these two distinct parts have similar properties to each other, and to the original cube? Suppose we now break these two halves further, and continue to do so, over and over again. Will these minute specks of sugar also continue to taste sweet and exhibit the other properties of the original cube? Can this decomposition proceed forever, or will we reach a limit that cannot be overcome? If we reach this limit, will the final minute particles still be sugar with all its properties, or will they be small specks of something else entirely?

What we have described above is a *Gedankenexperiment*, or thought experiment. These intellectual exercises are used frequently in philosophy,

Quantum Mechanics, and other branches of modern physics, to explore hypotheses in a logical fashion. Leucippus and Democritus, two philosophers from the 5th Century BCE, were probably the first to suggest that if matter is broken down as outlined above, eventually one will arrive at small particles that are indivisible [3]. The name "atoms" was coined from the Greek "*atomos*", meaning indivisible.

Democritus held the belief that atoms moved freely in a void, colliding with each other, and sometimes sticking together. It was a remarkably accurate picture of the modern viewpoint, considering its antiquity. However, the atomic theory was rejected by Aristotle, certainly the most influential philosopher of Ancient Greece (whom we met in Chap. 1), and so was disregarded for two millennia, up until the nineteenth century. Paradoxically, science is sometimes best served by a healthy disrespect for authority.

Further development of the concept of atoms had to await the work of John Dalton (1766–1844), an English chemist with a Quaker background. His atomic theory, which underpins modern chemistry, held that all matter is composed of atoms, which are indivisible and indestructible. The simplest materials, known as *elements*, are comprised of identical atoms. More complex materials, or *compounds*, are made up from combinations of different types of atoms. Chemical reactions proceed by rearranging the atoms of different compounds: each new arrangement corresponds to a new compound with chemical properties differing from those of its antecedents.

Dalton's theory was successful at explaining much of chemistry, in particular the fact that elements combine in particular ratios to form compounds. Ludwig Boltzmann in the late 1800s expanded Dalton's ideas to explain the temperature and pressure of a gas as a consequence of the motion of the constituent molecules. However, in the eyes of many 19th Century physicists, the atomic theory suffered from one overriding disadvantage: the atoms are so small that surely they must be completely unobservable. As we remarked in Chap. 3, a physical assumption must be *falsifiable*: if a quantity cannot be observed, it is outside the scope of physics.

This fundamentally philosophical disagreement on the nature of matter was resolved by Albert Einstein in the first of his ground-breaking papers of 1905 [4]. In 1827, a botanist, Robert Brown, had observed that tiny particles in a fluid, such as air or water, exhibited a jerky, random motion. This *Brownian* motion, named after its discoverer, was suspected of being the result of collisions of the fluid's molecules against the particles. If a coloured dye were added to water, the molecular collisions produced a slow diffusion of the colour throughout the liquid. Einstein studied this problem mathematically, and found that the diffusion rate could be related to the size of the molecules. The precision of his results convinced the doubters, and atoms became accepted not just as a mathematical artifice to explain chemical reactions, but as real objects. The unobservable had become observed, albeit indirectly.

But let us return now to the debate, which lasted several centuries, about the nature of light. As we have seen, Maxwell's theory seemed to have put an end to the dispute, but there remained an unexplained but important phenomenon called *black body radiation* (see Appendix 5.2). Suppose that we carry out an experiment by placing a small piece of iron in a blacksmith's forge and observe the change as we slowly turn up the heat. At first the iron starts to glow a dull cherry red, but as it gets hotter, its red colour becomes lighter and more yellow. If we can make it hot enough, the iron will glow with a white light, and if it could be made as hot as the inside of some stars, it would look blue.

This is a commonly observed effect that everybody who has lived in the era of tungsten filament lamps has witnessed, but for decades physicists strived in vain to explain it quantitatively. All their calculations predicted that most of the radiation would be emitted at ultraviolet frequencies, and they dubbed the phenomenon: *the Ultraviolet Catastrophe*.

It was not until nearly four decades after Maxwell's work that Max Planck in 1900 produced an explanation of black body radiation. He made the radical and, at the time, arbitrary assumption that electromagnetic radiation was not emitted continuously, but instead came out in discrete *packets*, or *quanta*. There was no justification for this assumption, which was completely contrary to the tenets of classical physics (see Appendix 5.3), other than that it worked! Planck proposed that the energy of these packets is directly proportional to the frequency of the electromagnetic radiation. The constant of proportionality was given the symbol h, which is now known as Planck's constant.

Another challenge for Maxwell's theory arose in 1887 when Heinrich Hertz investigated the photoelectric effect, which consists of the emission of electrons when light is allowed to fall on a metal surface. Hertz discovered that the energy of the emitted electrons depends on the frequency of the light, and not on the beam intensity, as expected. In 1905, Einstein explained the anomaly by proposing that the incident light beam is not a continuous wave, but rather is composed of discrete *wave packets* (now called photons).[4]

---

[4]It may be worth recalling that Einstein obtained his Nobel Prize in 1921 for the photoelectric effect, not for his work on Relativity, since the foundation statute required that prizes could be awarded only for research validated by experimental confirmation, which was not yet the case for General Relativity.

The energy of each photon is proportional to its frequency, and the photons can be identified with the quanta of Planck.

Readers will by now be querying why it is so difficult for physicists to distinguish between waves and particles. Surely the two phenomena are nothing like each other. We've all seen ripples on a pond, we've all held a pebble in our hand. So, what's the problem? Is light a form of electromagnetic waves, or is it made up of photons shot out like bullets?

The answer, however, is not as simple as the question: sometimes light is one thing, sometimes the other. Being a truly Pirandellian creature, it has a third, more complex nature, in which both the wave and particle personalities coexist. We will try to explain this duality in the following Sections.

5.4   The QM Revolution

We have already commented on the scarcity of Common Sense in the world at large. The Italian poet, Giuseppe Giusti (1809–1850), has written an epigram to denounce sarcastically what he perceived as the disappearance of *Common Sense* in the school room:

Il Buonsenso, che già fu caposcuola,	Common Sense, who used to be
ora in parecchie scuole è morto affatto;	school master,
la Scienza sua figliuola	is now in many schools dead and
l'uccise, per veder com'era fatto	buried;
	the reason is that Science, who is his
	daughter,
	killed him to see how he was made

Now that we are exploring Quantum Mechanics, we have to do precisely the opposite to what Giusti was advocating, i.e. we must learn to abandon common sense in favour of experimental evidence. Otherwise we will never *understand* how an object such as a photon, can be simultaneously both a particle and a wave. As we learned in Chap. 2, to *understand* sometimes means simply to accept a premise and explore its consequences. This was the approach adopted by Max Planck with his "explanation" of black body radiation.

This trait appears to be much easier for a young, uncluttered mind. So it happened that a Ph.D. student, Louis-Victor Pierre Raymond (Louis) de Broglie, with a thesis of just 70 pages [5], obtained both his Ph.D and the Nobel prize (in 1929), by postulating a dual nature, not only for photons, but also for matter. (Before this, he had already obtained a degree in history and one in science.) De Broglie's bold assumption meant that electrons, the so-called atoms of electricity, should also display wave-like characteristics, such as diffraction and interference, which we discussed in Chap. 1 for light. This prediction was confirmed by the experiments of G. P. Thomson and of Davisson and Germer (1923–1927), who diffracted a beam of electrons from a nickel crystal.

But how can we depict a particle with a dual nature? Classically we describe a particle's location by specifying its coordinates in four dimensions (4D)— the three spatial dimensions and time. In the case of wave motion, we normally specify the *amplitude* of the wave in terms of these 4D coordinates. The physical interpretation of *amplitude* depends on the type of wave we are considering and of the medium supporting it. For instance, in the case of a sound wave, amplitude represents the displacement of the medium through which the wave is propagating (usually air). In the case of light and other forms of electromagnetic radiation propagating through a vacuum, there is no physical supporting medium, and the amplitude represents the value of the electric (or magnetic) field at these 4D coordinate points, bearing in mind that the amplitude may be either positive or negative. The intensity of the light is given by the square of the amplitude, ignoring the sign.

Now we see the conundrum that presented itself to physicists in the early twentieth century. Common Sense will get us nowhere. Instead we must abandon ourselves to mathematics, and see where it leads. The modern interpretation is that associated intimately with every particle is a *wave function*. This wave function is a mathematical abstraction, and as far as we know has no physical reality. Indeed, it is a *complex* quantity, which in mathematics means that it has both *real* and *imaginary* parts. (As we have seen in Sect. 4, an *imaginary* number comes from taking the square root of a negative number.) If the reader is getting the sinking feeling that we are about to lose ourselves in an unintelligible quagmire, please take heart, for we shall go no further in this direction.

Clearly the wave function must have some connection with reality, or it would remain an irrelevant mathematical oddity. The connection comes via its *modulus squared*.[5] Not just electrons, but all matter, has wave-like properties. However, the interference effects are easier to observe with less massive particles. In analogy with the intensity of light described above, the modulus squared of the wave function at a particular point in space and time gives the *probability* of finding the particle to be located at that point at that time.

---

[5]The modulus (or absolute value) of a real quantity is its magnitude, ignoring whether it is positive or negative. However, wave functions are *complex* quantities, which complicates matters slightly. The modulus squared of a complex quantity is the sum of the squares of the real and imaginary parts.

This last sentence contains the very core of what is now called Quantum Mechanics.

What we have said above is truly revolutionary: we can no longer say that a particle is *here* or *there*, but only specify the probability of finding it here or there. Not only that, if the particle is *there*, in order to *look at it*, or find a way to somehow detect it, we must use a small amount of energy (light waves have energy) and this energy moves the particle somewhere else. In other words, the process of observation changes, if ever so slightly, what we have observed.

Gone is the *determinism* of classical physics, where it was believed that if one could specify the positions and velocities of all the particles in the universe, one could predict the future for all eternity,[6] and as an extra bonus also reconstruct the past. What takes its place is a world where one can predict nothing with complete certainty, but only estimate the probability that some event will take place. With macroscopic objects, the uncertainties are negligibly small: if one leaps off a tall building there is very little chance, a second or two later, of finding oneself reposing in an armchair reading Newton's *Principia*. In the domain of atoms, however, analogous phenomena are entirely possible.

Having now grasped some idea of the relevance of the wave function, the question arises: how does one calculate it? The answer was provided by Erwin Schrödinger in 1926, who proposed an equation that bears his name. The solution of Schrödinger's Equation for a particle in a particular environment, e.g. an electron in the hydrogen atom, provides the wave function for that particle. This wave function can be used to calculate essentially everything associated with the particle, i.e. its position, velocity, orbital angular momentum, etc. However, these quantities are obtained and expressed only as probabilities.

In our everyday world, we are used to being able to predict events with near certainty. Astronomers can tell you when solar eclipses will occur hundreds of years in the future. So what use is it if the best we can get from QM is a probability? It is like being told that there is a 1 in 36 chance of rolling two sixes on a pair of dice. It doesn't help decide whether to bet your hard-earned cash on the next roll or the one after. (In fact, one shouldn't waste any time on this decision. If the dice are "fair", the chance of rolling two sixes is the same for every throw, irrespective of past history.)

There are, however, many instances where probabilistic predictions are rather more useful. Weather forecasts are not exact, but most people consult them when planning a picnic. Casinos know very accurately the likelihoods of

---

[6]This idea was first put forward by Pierre-Simon Laplace in 1814, and is known as Laplace's demon (see Appendix 5.4). It has been debated by physicists ever since.

a gambler winning on their various poker machines. They plan their budgets based on this knowledge. When a large number of gamblers use the machines, the reality reflects these predictions accurately. Actuaries in insurance companies examine past records to determine the risk associated with the various behaviours and lifestyles of their clients, and adjust the premiums they must pay accordingly. Some of the classical laws of physics, e.g. those dealing with the properties of gases, are statistically based. They achieve their accuracy because of the enormous numbers of molecules involved, which make the statistical predictions very precise indeed.

However, when dealing with individual atoms, QM does not allow an accurate prediction of events. In a sample of radium, one cannot tell in advance which will be the next atom to undergo radioactive decay, only how many are expected to decay in a given time.

5.5   Heisenberg's Uncertainty Principle

As we have just seen, if we wish to know the location of a particle, we need to obtain its wave function by solving Schrödinger's equation. We will not obtain a precise location, but only a probability distribution for the particle. If the location is known fairly accurately, this distribution will be narrow; contrarily, if the distribution is broad, it means that our uncertainty of the particle's location is great.

We could apply the same methodology to determining the particle's velocity (or more strictly, its momentum, i.e. its mass multiplied by its speed). Again, we would obtain a probability distribution, and a narrow one would mean we have a fairly precise knowledge of the momentum, and a broad one the opposite.

Heisenberg in 1927 realised that these two probability distributions, one for the particle's location, and the other for its momentum, are not independent of each other, but closely related. If we can locate a particle accurately, we can measure its momentum only roughly, and vice versa. In the extreme case, where we are able to locate the particle exactly, we can have no knowledge at all of its momentum.

Of course, in physics we do not deal with "hand-waving" statements, such as "accurately" or "roughly". Heisenberg presented his principle in a mathematical form, stating that the product of the *uncertainty* in the particle's location and the *uncertainty* in its momentum cannot be less than $\hbar/2$, where

$\hbar$ is the ubiquitous Planck's constant h divided by 2π. Here by "uncertainty" we mean the standard deviation,[7] as it is called in statistics.

Heisenberg's Uncertainty Principle is often explained by saying that any attempt to measure the position of a particle will perturb the particle's momentum. The more accurately we try to locate the particle, the greater is this effect. Conversely, any attempt to measure its momentum, dislodges the particle from its original position by an unknown amount.

The pair of quantities, "position" and "momentum", which satisfy Heisenberg's Uncertainty Principle in this manner, are called *canonical conjugates*. They are not the only such pairs that satisfy this relationship. Energy and Time are another example, as are Angular Position and Angular Momentum.[8] The latter are important in the physics of electrons bound within the confines of an atom.

In the everyday world where we live, the idea that one cannot know one's velocity and position at the same time seems strange indeed. Certainly, the traffic police have no such qualms when they send you an infringement notice for violation of the speed limit at a particular place and time. They are only able to justify this action because of the extremely small value of Planck's constant.[9]

5.6   Collapse of the Wave Function

We are now at a stage where we can explore some of the consequences of the probabilistic nature of QM, and provide an insight into the "incredible" in the title of this Chapter. It is time to put away the last remnants of our common sense, and follow our fancy and mathematics, wherever they may lead us.

Let us begin with a simple thought experiment, where we imagine shooting a beam of particles, e.g. electrons, through a small slit. There are various ways to construct such a slit, which need not concern us here. The important thing is that suitable slits can be constructed without much difficulty in the laboratory. The wave-like nature of electrons means that we expect diffraction to occur as the waves pass through the slit.

---

[7] For a bell-shaped (or "normal") distribution of data around a central point x, the standard deviation σ is defined such that 68% of the data will lie between $x - \sigma$ and $x + \sigma$.

[8] According to the story at the start of this Chapter, in Neutrinia "money" and "time" are also a pair of canonical conjugates.

[9] In international units (m, Kg and s) the value of $\hbar$ is approximately 10^{-34}.

As we have already learned earlier, diffraction is the intrusion of waves into the shadow regions behind obstacles, and is a basic phenomenon, well explained by Maxwell's theory, associated with all wave motion. In the case of waves passing through a slit, we expect most of the energy to pass into the region directly behind the slit, but a small amount will appear in lesser peaks in the shadow regions to either side of the main peak. This situation is shown graphically in Fig. 5.3. The spacing of the peaks depends on the width of the slit and the wavelength of the incident waves.

In the case of the diffraction of light, if we placed a photographic film at a distance behind the slit, we would see a dark splodge where the main beam lands on the film, and lighter splodges from the diffracted beams on either side. If we narrow the slit, the diffracted beams would extend further out on either side. Those with an interest in photography will recall that decreasing the aperture of a camera lens (i.e. increasing the f-number) results in sharper images at first due to less sensitivity to lens aberrations, but then the fuzziness increases when diffraction effects start to become significant.

Returning now to the case of electron diffraction, if we place a number of electron-detectors across the region behind the slit, we can count the number of electrons arriving per second in the detectors at the various angles. This number will be greatest in the detector directly behind the slit, but the smaller diffraction peaks will also be readily discernible by the clicks from the detectors at angles corresponding to these peaks.

At the moment when one of our detectors emits a click, we know immediately where that particular electron is located: it is located inside that detector,

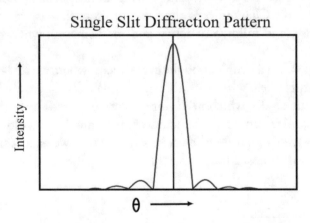

Single Slit Diffraction Pattern

Fig. 5.3 The diffraction pattern from waves passing through a single slit as a function of the angle of diffraction θ

and there is zero probability that it is located elsewhere. Our act of measurement has resulted in the wave function of the electron shrinking to the size of the detector. In QM this is known as the *collapse of the wave function*.

To reiterate, up until the moment of detection we had no idea exactly where the electron was located; instead our knowledge was confined to a probability distribution that we obtained from the electron's wave function, showing the statistical likelihood of it arriving in each of the various detectors. After the moment of detection, we are now certain where the electron is, and also, *simultaneously*, we are certain of where it is not, even if the other detectors are far away from the first one. In the next Chapter, however, we will learn that such a simultaneity is in conflict with Einstein's Theory of Relativity. This conflict remains one of the outstanding problems of modern physics. We will discuss it in more detail in Chap. 6.

The issue of the collapse of the wave function caused by observation is itself a controversial topic, and has been debated at length by philosophers and physicists alike. The impression that the *observer* triggers the collapse by the act of measurement attaches an anthropomorphic element to QM. An extreme of this approach is the *Many Worlds Interpretation* of Hugh Everett, who proposed in a Ph.D. thesis at Princeton in 1973 that the universe splits into parallel divergent universes whenever a measurement is taken, each of them bound to the *choice of detector* made by the arriving particle. This proliferation of universes continues as time progresses, until essentially, somewhere at some time, there exists a universe where anything we can imagine is possible. (Where is Occam's Razor when you most need it?).

Part of the difficulty may lie with the tendency to interpret the wave function as possessing a physical reality. Rather, it should be viewed as a mathematical artifice that enables the calculation of the probable location of the electron. If the electron is suddenly localised because of an experiment providing extra information (e.g. that the electron has been found in detector No. 1), then clearly the probability distribution of the electron's location needs to be updated (as there is now no probability of the electron being located in other detectors). This is what we mean when we say that the electron's wave function has collapsed.

An intriguing puzzle that has been known to fool even professional mathematicians may cast a little light on the above situation. Known as the Monty Hall problem after an American TV games show host, it runs like this: *Contestants stand before three closed doors. Behind one of the doors is a car, while the other two conceal booby prizes. The contestants choose one of the doors, hoping to win themselves the car. The show host, who knows where the car is, does not open the contestant's choice, but opens instead one of the two other doors, revealing a booby prize. He then offers the contestants a choice of staying with their original selection, or changing their mind and selecting the other remaining closed door. What should they do?*

The *intuitive*, or *common sense*, answer is that it doesn't matter, as there are now only two closed doors, and surely there is an equal likelihood of finding the car lying behind each of them. This answer is *wrong*. Contestants will have *twice as much chance* of winning the car if they change their selection from their original choice to the other door.

The solution of the Monty Hall problem is discussed in Appendix 5.5. However, the point we wish to make here is that as soon as the host opens the door revealing the booby prize, the probability distribution associated with the car's location is changed. Originally there was an equal probability of the car being located behind each of the three doors. Now we know for certain that it is not behind the door that was opened. We have received extra *information* that changes the probability distribution of the car at this moment. *Information* is an important concept in QM, but one which we cannot pursue further here.

5.7   Wheeler's Delayed Choice Experiment

Let us continue with our thought experiments by envisioning two narrow slits close together and allowing a plane wave (in the wave representation), or a stream of particles (in the corpuscular representation), to be incident upon them. The setup is sketched in Fig. 5.4. We will discuss the case of light waves (i.e. photons), but we could just as well have used electrons, or any other quantum particles.

From what we have already discussed, we would expect the waves passing through the two slits to interfere, and produce an interference pattern of light and dark fringes on the optical screen placed behind the slits similar to that shown in Fig. 5.5.

Naïvely, we may tell ourselves that one photon has passed through one slit, a second through the other slit, and their wave functions have interfered with each other to produce the above interference pattern. However, it is not quite so simple, and QM has a few more tricks to play.

Let us continue our thought experiment by reducing the intensity of the incident beam so that only one photon (or electron, if we wish) is in flight at any time. In other words, we wait until one photon has collided with our photographic plate, before releasing the next one on its journey. When the photon that is underway reaches the slits, it should pass through one or the

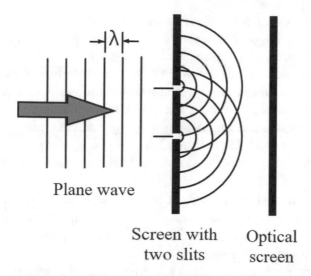

Plane wave

Screen with   Optical
two slits   screen

Fig. 5.4 Light waves passing through two slits and impinging on an optical screen, or a photographic plate

Fig. 5.5 Interference pattern from light (coherent) passing through two slits and falling on a photographic plate. Image, public domain, courtesy of Pieter Kuiper (Image adapted from https://commons.wikimedia.org/wiki/File:SodiumD_four_double_slits.jpg (accessed 2020/05/23))

other, be diffracted in the manner we discussed in the last Section, and then impinge on the photographic plate. We would expect over time to see a build-up of the diffraction pattern from one slit (Fig. 5.3) superimposed on that of the other, but not the fine interference structure shown in Fig. 5.5. However, the observed interference pattern remains exactly the same as in Fig. 5.5. It just takes longer to build up because of the weak beam intensity.

It is as though the photon is not really specified in location until it has impinged on the photographic plate, and its wave function has collapsed. Before then it could be anywhere, and it could be passing through either slit, or even through both slits simultaneously, such is its ghostly nature.

Okay, you say, then let us see if we can find a way to detect which slit it is passing through. Suppose we remove the photographic plate, and replace it with two telescopes, one directed at the first slit and the other at the second. If a photon passes through the first slit, a flash of light will be detected by the first telescope; vice versa, a photon passing through the second slit will be detected by the second telescope. We keep the beam intensity low enough to ensure that only one photon is in flight at any particular moment. With this setup, we are successful. Half of the photons are observed to pass through the first slit and half through the second, exactly as if we were looking at a stream of bullets.

In other words, whether the light beam behaves as a wave or a stream of particles seems to depend on what apparatus we are placing in its passage to detect it. Any attempt to localise which slit the photon passes through destroys the interference pattern.

There is one further twist we can add to our thought experiment. Let us set up an experiment similar to that of Fig. 5.4, but construct it so that we can randomly choose whether to use a photographic plate (i.e. a wave detector) or telescopes (photon detectors), and are able to replace one choice by the other very quickly. We also lengthen the time of flight between the slits and the detectors so that there is time for us to make our choice *after* the photon has already passed through the slit(s) and is on its way to the detectors.

This experiment, which is known as the *Wheeler Delayed Choice Experiment*, has been performed for a variety of quantum particles (some as large as whole atoms [6]) with results that destroy the last vestiges of Common Sense. If we observe the slits with a telescope, we can detect which slit the particle passed through; if we put the photographic plate in place, an interference pattern will build up, which is indicative that the particle passed through both slits. However, the particle had already passed through the slit(s) *before* we made our decision which detector to use. How did the wave/particle know which choice we were going to make before we knew it ourselves?

O, that way madness lies; let me shun that; No more of that [7].

We shall discuss the Wheeler Delayed Choice Experiment further in Chap. 12 (Part 3).

5.8   Quantum Entanglement

If we are still reeling from the implications of Wheeler's Delayed Choice Experiment, then "*a cup of tea, a Bex, and a good lie down*"[10] are surely necessary before tackling the mysteries of *Quantum Entanglement*.

From our examples above, one may obtain the impression that everything in QM is of a ghostly probabilistic nature, and that nothing can be known for certain. As we have seen, at the microscopic level this is largely true, but there are some quantities that must not change, and these are expressed by *conservation laws*. These laws are the quantum analogues of corresponding laws in classical physics. Indeed, they are more than analogues: they are really the same laws, with the quantum form transitioning into the classical form as the objects we are considering increase in complexity and size. Examples are the Laws of Conservation of Momentum, Angular Momentum and Energy. We shall discuss conservation laws in more detail in Chap. 8.

Let us consider another thought experiment where two electrons are produced together such that their total combined angular momentum is zero. Such a process is possible if the electrons are produced jointly in a physical process where the total angular momentum that the pair can carry away is limited to zero by the requirement of the Law of Conservation of Angular Momentum. Then if one electron has a positive angular momentum, the other electron must carry an equal *negative* angular momentum. In other words, the two electrons are spinning at the same rate in opposite directions. This is a classical picture—electrons are *not* simply spinning balls, as we shall see in Chaps. 8 and 9. However, this model will suffice for our purposes here. Two particles restrained in this manner by a conservation law are said to be *entangled*.

It is worth reiterating that at this point we have not yet conducted any measurements on the electrons, so we cannot say what the angular momentum of each electron is, only that they spin in opposite directions, and that the sum of their individual angular momenta must be zero. Indeed, QM goes further and states that until a measurement is made, neither electron has a well-defined spin, but is a mixture, or superposition, of possible spin states.

Let us now conduct our measurement. We select one electron and measure its angular momentum, while we allow the second one to continue on its

---

[10]A marketing slogan for Bex, a widely used but addictive analgesic containing phenacetin. Bex was banned in 1977, when it was shown to cause renal cancer. The slogan became part of the Australian vernacular.

journey. Our act of measurement collapses the wave function of the first electron, resulting in its angular momentum becoming precisely defined (i.e., no longer described by a probability distribution). However, because the two particles are entangled, the wave function of the second electron must also *simultaneously* collapse, and its angular momentum becomes precisely specified, *no matter how far away the second electron has moved from the first*.

This aspect of QM was not accepted by Einstein, who maintained that this "*spooky action at a distance*" violated the Theory of Relativity, which indeed it does (as we shall see in Chap. 6). He claimed that QM must therefore be incomplete, and that the electron spins were indeed well defined, although hidden from experimenters until they had carried out their experiments. This interpretation was known as the "*hidden variables*" theory. A major contribution by the Irish physicist, John Stewart Bell, was to show that no hidden variables theory can ever reproduce all the predictions of QM.

The first experimental test of the Einstein versus Bell disagreement was performed in 1972, with the evidence strongly in favour of Bell [8]. Since then, many tests have been made to close possible loopholes in the methodology. As time has passed, the experimental evidence in favour of entanglement has accumulated, and now few physicists doubt its existence. Many research projects are underway to harness its capabilities in such diverse fields as Quantum Computing, cryptography, ultra-precise clocks and biology.

5.9   Schrödinger, and His Long-Suffering Cat

No Chapter on QM would be complete without a mention of Schrödinger's cat. The story has been told so often that it has become hackneyed, so we apologise in advance if the reader is already familiar with it. However, it is a thought experiment that raises important questions about the boundaries between quantum and classical physics, and is worth presenting for that reason. There are several variants to this experiment, but the one below is a typical example.

Imagine that we have a box containing a cat and a weakly radioactive source that has a probability of 50% of emitting one radioactive particle per hour. Also in the box we have a Geiger counter to detect the decay product from the source. The Geiger counter is wired so that when it registers a detection, it releases a hammer which swings down and smashes a phial of cyanide gas, thereby releasing poisonous fumes into the box and killing the cat. Remember, this is only a thought experiment, and not intended to be carried out![11]

After one hour, when the experimenter opens the box there are two possibilities:

(1) The radioactive source has not emitted a particle, the cyanide has not been released, and the cat is alive and well;
(2) The radioactive source has emitted a particle, the cyanide has been released, and the cat is dead.

From a quantum perspective, one might say that it is the observation process of opening the box that collapses the wave function and reveals which of the two possibilities is real. Before the box is opened, both possibilities exist and the cat is in a mixture, or superposition, of two states, one in which it is alive and one in which it is dead.

From a classical physics viewpoint, one would say that the cat is alive or dead, but until we open the box and take a look, we do not know which. This is akin to Einstein's Hidden Variables theory that we discussed in the previous Section. So under what conditions are the classical theories of physics accurate, and when do they fail, so that we must turn to QM? This is the dilemma that Schrödinger's Cat has highlighted for us.

Macroscopic objects, such as pebbles, balls, cats, and even us, contain countless billions of molecules.[12] The wave function associated with these large objects is a linear superposition of the wave functions associated with the atoms of which they are constructed. For two waves to interfere, there must be a stable phase relationship between them. This means that the peaks and valleys of the waves are not randomly distributed. In this case, the waves are said to be *coherent*. In a large body, the wave functions associated with the billions of molecules are jumbled, and so do not add up coherently. In this case, the superposition of the wave functions produces physical effects that are just averages of the individual effects. This is akin to the classically predicted behaviour.

As an example, the light emitted by a tungsten filament lamp is incoherent, and so will not normally produce interference effects.[13] On the other hand, light from a laser is produced by stimulating the atoms in the laser to all

---

[11] Our satirical spoof, *Doppelbelcher and the OOO*, which opens Chap. 12, illustrates some of the disastrous consequences of trying to implement Schrödinger's thought experiment in the laboratory.

[12] In the vicinity of 10^{23} molecules per kilogram.

[13] Interference effects, e.g. Young's experiment, can be carried out with an incandescent filament lamp if a narrow slit is placed in front of the filament to block out most of the light. Only light from a relatively small number of atoms is then allowed to pass through.

emit simultaneously. This light is coherent, and therefore readily produces interference effects.

In the above experiment, the cat is clearly not a quantum particle, and the radioactive particle that triggered the Geiger counter just as clearly is. The boundary between the classical and quantum domains is continually being raised. In 2010, Andrew Cleland and his team at the University of California, succeeded in placing a "paddle" 30 microns long containing trillions of atoms into a quantum state [9]. By cooling the paddle below 0.1° K, they placed it into its "ground" state, where the available thermal energy is too small to allow any excitations to higher energy states to occur. They then added the smallest possible unit of vibrational energy. This created a situation where the paddle was in a superposition of two states, one with zero energy and the other with one unit of vibrational energy. *"This is analogous to Schrödinger's cat being dead and alive at the same time,"* said Cleland. When the experimenters measured the energy, the wave function collapsed, and the paddle had to "choose" whether to remain in the excited state, or pass its energy to the measuring device. The paddle was large enough to be visible.

There are a number of interpretations of the Schrödinger Cat experiment, some of which involve live and dead cats in various universes. The philosophical implications of the thought experiment are profound, involving as they do a live animal, and notions of life and death. Let us consider for a moment replacing Schrödinger's cat with a virus. Depending on one's definition of *life,* a virus is considered by many to be a living creature. Viruses range in size from 0.004 to 0.1 microns, and are thus much smaller than Cleland's paddle. Quantum effects should surely apply to them. Bacterial cells range from about 1 to 10 microns in length and from 0.2 to 1 micron in width. Bacteria are most certainly alive, and are also smaller than the paddle, so we can expect quantum effects to be relevant to them. We begin to see some of the challenges thrown to philosophers by Quantum Mechanics.

In the next Chapter, we shall present the next great challenge to *Common Sense*, and the other half of the revolution in physics that took place in the twentieth Century, namely, the Theory of Relativity.

References

1. Maxwell JC (1865) A dynamical theory of the electromagnetic field. Royal Society, Great Britain
2. Pirandello L (1933) Uno, Nessuno, Centomila One, None, and a Hundred Thousand, Trans. English by S Putnam (1933), published on Project Gutenberg. Accessed 22 July 2020

3. Democritus, c460 BCE - c370 BCE Nothing exists except atoms and empty space; everything else is opinion
4. Einstein A (1905) Über die von der molekularkinetischen Theorie der Wärme geforderte Bewegung von in ruhenden Flüssigkeiten suspendierten Teilchen (About the movement of particles suspended in quiescent liquids required by the molecular-kinetic theory of heat) Annalen der Physik 17(8): 549–560
5. de Broglie L (1924) Recherches sur la Théorie des Quanta. Ph.D thesis: Research on Quantum Theory
6. Jacques V et al (2007) Experimental realization of wheeler's delayed-choice Gedanken experiment. Science 315(5814):966–968
7. Shakespeare W, King Lear, Act 3, scene 4, 17–22
8. Freedman SJ, Clauser JF (1972) Experimental test of local hidden-variable theories. Phys Rev Lett 28:938
9. A. D. O'Connell et al (2010) Quantum ground state and single-phonon control of a mechanical resonator. Nature 464:697–703. https://www.nature.com/articles/nature08967. Accessed 23 May 2020

6

Special Relativity

There Was a Young Lady Named Bright
Whose Speed Was Far Faster Than Light.
She Set Out One day
In a Relative Way
And Returned on the Previous Night.
— A. H. Reginald Buller [1]

6.1 A Vietnamese Fairy Tale: The Land of Bliss

Over five centuries ago, a young mandarin, named Tu Thuc, was chief of the Tien Du district in what is now Vietnam. He was a devoted scholar and possessed many books on all subjects, except for any on the Land of Bliss, which is known to be a fairy place and a land of eternal youth and pleasure.

Once when travelling during the Flower Festival, Tu Thuc came across a disturbance near an ancient pagoda covered in beautiful blossoms. A young maiden of startling beauty had just grasped a branch to better admire a sacred moon flower, when the branch had broken off in her hand. The priests arrested her and imposed a heavy fine, which she was unable to pay. Although Tu Thuc had insufficient money with him to cover the fine, he offered the priests his expensive brocaded coat, which

© The Author(s), under exclusive license
to Springer Nature Switzerland AG 2021
R. Barrett and P. P. Delsanto, *Don't Be Afraid of Physics*,
https://doi.org/10.1007/978-3-030-63409-4_6

they accepted as payment. Tu Thuc was praised by the local community for his generosity.

Some years later, tired from the duties of his office, Tu Thuc resigned and set out to explore his native land and write poetry on events and scenes that inspired him. High in the mountains, he sat down to rest and composed a poem on the Land of Bliss (see Appendix 6.2). As he completed it, a cavern appeared before him in a mountain, and it seemed almost to beckon him. He entered and passed into a land of crystalline springs, lotus leaves and wonderful birds with bright colourful feathers. A choir of angelic voices sang gently in the background.

A group of lovely maidens presented him to their queen, who smiled and welcomed him. He was astonished to see at her side the young maid he had befriended years before. "This is my daughter, Giang Huong," she said. "We have never forgotten the help you gave her when she was in distress." Tu Thuc felt his heart go out to the demure fairy princess. "You are obviously a kind and generous man," continued the queen, "and I offer you my daughter's hand in marriage."

The wedding was celebrated that day, with great pomp and festivities. Tu Thuc had never been so happy. However, as time passed, he began to feel nostalgia for his old village. Finally, he asked the Queen if he might be allowed to visit his friends and relatives for just a few weeks. He promised to return, for his love for his young wife was as strong as ever.

The queen reluctantly acceded and suggested to Tu Thuc that he lie down on the ground and close his eyes. He did so and fell asleep immediately. When he awoke, he looked around but could not recognize where he was. He stopped a passing old man and said to him. "I am Tu Thuc, and I am seeking my village in Tien Du."

The old man shook his head. "There was once a Tu Thuc, who was chief of the Tien Du district, but that was a hundred years ago, long before I was born. He set out on a long journey, and never returned. Some say he perished, and others that he found a secret way into heaven."

Tu Thuc realized that during the time he had been in the Land of Bliss, a hundred years had passed by on earth. There was nothing left for him anymore in the village of his birth. All his friends were dead, and the customs of the young people appeared strange to him. He set out once more in the direction of the mountains and was never seen again. Whether he found his way back to the Land of Bliss, or became lost and died in the rugged highlands, is not known.

6.2   What is Time?

The above tale is a traditional Vietnamese fairy story. In the manner of such stories, there are several variants which differ in their details and endings. It has now been adapted into "*The Vietnamese Legend, Giang Huong – the Musical*", which has been staged in the Opera House of Ho Chi Minh City.

A feature of the story is that in Fairyland, time flows more slowly than on earth. This is a characteristic shared with other tales. "*Rip Van Winkle*" by Washington Irving is one of the best known, where Rip falls asleep and wakes to find twenty years have passed him by. In "*The king of Elfland's Daughter*" by Irish writer, Lord Dunsany, there is much in common with the Vietnamese story.

However, these are only stories, where the writer can create whatever fantasy he or she desires. In the real world, here on earth, time has no surprises, but passes at the same rate for everybody, wherever and whenever they are.

Or does it?

Before we can talk of time, and whether it races or crawls by, we need to be clear exactly what we are talking about. The Oxford English Dictionary offers the following definition for time: "*the indefinite continued progress of existence and events in the past, present and future regarded as a whole.*" This may be all right, so far as it goes, provided that one does not probe too deeply into what is meant by past, present and future. For instance, past is defined as "*gone by in time and no longer existing*", and the circularity of the definition becomes immediately apparent.

St Augustine (354–430 CE) once famously said: "*What then is time? If no one asks me, I know what it is. If I wish to explain it to him who asks, I do not know.*" Since then, philosophers have debated the subjectivity of time, and particularly *duration* which is a measure of elapsed time. The difficulty arises because the only time one can actually experience is the present, and an estimate of elapsed time requires the memory of a past event, which is subjective. Time appears to pass more quickly if we are having fun.

Physicists have long tried to put the measurement of time on a quantitative basis. Galileo, in his experiments on the mechanics of falling bodies relied on his own pulse, and a primitive water clock, to measure the elapsed time. He later studied pendulums, which led to the development in 1656 of a pendulum clock by Christiaan Huygens. The construction of the chronometer in the Eighteenth Century by John Harrison enabled the accurate determination of longitude by ships at sea, and revolutionized the art of navigation. Today, the most accurate clocks use the period of electromagnetic radiation emitted from the Caesium-133 atom as the standard unit of time.

Before the advent of train travel and timetables, village life was leisurely, and it mattered little if the clocks in one village were running slightly ahead or behind those of another village further down the road. Now, with data exchanged around the globe at the speed of light, much more importance is attached to time standards. Surely, with today's incredibly accurate clocks, we would notice if time in one place were passing at a different rate from time somewhere else.

Or would we? We shall see.

It has long been the custom of physicists to designate the position of an object in the three-dimensional (3D) space in which we live by three coordinates. We begin by choosing a reference point in space, from which all other dimensions are measured. This point is known as the *origin*, and is designated by the letter O. It can be anywhere, but it is more practical to choose an origin that is convenient for the problem we are studying. Choosing an origin in another star system, such as Alpha Centauri[1] is not appropriate for calculating the location of planets in our own solar system.

Next, we choose a direction emanating from the origin. Again, this is arbitrary, and is usually called the *X axis*. Two other axes are chosen at right angles to the X axis, and to each other. These are the *Y* and *Z axes*. Figure 6.1 makes all this clear. The position of any point P in 3D space is denoted by x, y, and z, which are called the coordinates of the point, and are the distances in the X, Y, and Z directions, that one must travel from the origin to arrive at the point P.

So far, so good. The three spatial coordinates are all that is required to define the position of a point in 3D space, provided the point is stationary in time. If the point is moving, then at each time t the point will have different values of x, y, and z. To describe such an evolving system, physicists introduce a fourth axis, called the T (time) axis, assumed to be perpendicular to the other three. An arbitrary time is selected for the origin of the T axis. The coordinate t is then the elapsed time since t = 0, which occurs at the origin.

---

[1]Alpha Centauri is the closest star system to us (about 4 light years distant) and one of the most luminous in our sky.

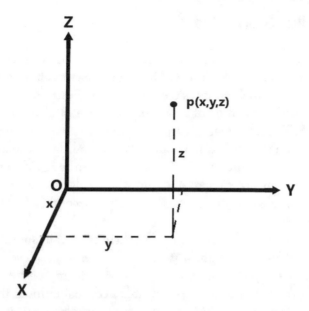

Fig. 6.1 Coordinates of point P in a 3D Cartesian coordinate system

This is the stage at which our intuition (or common sense) warns us that we have departed from reality. Space, as we know it, has only three dimensions, so how can we talk about a fourth axis perpendicular to X, Y, and Z? Nevertheless let us consider this four-dimensional (4D) *space–time* as just a mathematical device that enables us to formulate physical problems, e.g. the calculation of the trajectories of moving objects, in simpler mathematics. So now, in the same way that x, y and z identify a point in space, the four coordinates x, y, z, and t identify a point (which we shall call an *event*) in four-dimensional (4D) space–time.

As we have seen, there is an arbitrariness to the values of these four coordinates because of the arbitrary choice of the origin O. Peter might choose the location and time of his birth for his choice of O, while Mary might choose something altogether different. A clear requirement of any physical theory is that it must predict the results of experiments and observations, regardless of the assumptions made. Clearly the path of the moon around the earth does not depend on where or when Peter was born. Indeed, if our theory is good, we should be able to choose any origin we like, and still accurately predict the results of the measurements we are making. Let us explore this a little further in the next Section.

6.3   Walking on a Train

We all feel that we can distinguish when we are in motion from when we are at rest. However, the work of Galileo and Newton showed that making such a decision is actually impossible, except when our velocity changes. At the moment, sitting in front of a computer screen, and reaching for a cup of delicious Ethiopian Harar coffee, I do not have any feeling of motion. However, I live on the surface of a planet that is rotating at a speed of roughly 460 m per second (at the equator). Also the earth is revolving around the sun at about 30 km per second, and the whole solar system is plummeting towards the constellation of Leo at 371 km per second. Would anyone call this "being at rest"?

The reason that I have no perception of these enormous speeds is explained by Newton's first law of motion: *every body continues in a state of rest or uniform motion unless acted upon by an external force.* I, and my coffee cup, and my computer, continue on our journey towards Leo, maintaining the same relative separation from each other unless some force acts upon one of us, e.g. if I inadvertently knock the cup and spill the coffee. Suppose I move to the study I have set up in my private jet aircraft.[2] Once again, after the plane has taken off and reached cruising altitude, I will have no sensation of motion, unless we encounter turbulence, which is an external force likely to send my coffee mug sliding across the desk.

Can we all, as young children, remember sitting in a train at a railway station, waiting impatiently for our journey to begin. Looking out of a window, we notice that at last we have begun to edge forwards and are finally underway. We sit back in our seat, turn towards our parents to tell them, and are astonished to see that the station platform on the other side of the train remains absolutely still. What we had observed was another train pulling into the platform on the track next to us and have mistaken its movement for our own.

Technically we can summarize the above considerations by stating that any system moving at constant velocity (which we call an *inertial frame of reference*) is equivalent to any other, in the sense that any experiment performed in one gives the same result if it is performed in another. Consequently, it is impossible to decide which frame of reference is moving, and with what velocity. Of course, if the speed of the train is not constant (i.e. it is accelerating or braking) we feel the effect of the acceleration or deceleration, but not

---

[2]We are all allowed our cosy private dreams.

of the speed. For example, if the train suddenly stops, we are thrown forwards and might even fall, if we happen to be on our feet at the time.

Suppose that Peter is standing on a platform, and Mary is on a train departing slowly from the station. Peter may claim that Mary is moving away from him at 10 kph, which he determines to be the speed of the train, as measured in his frame of reference. Mary, on the other hand, may claim equally validly that Peter is moving away from her at 10 kph in the opposite direction, and that she is the one at rest. If Peter throws a ball towards Mary at 20 kph, as measured by him, Mary will see it go past her at 10 kph. This is all quite straightforward and is explained readily by Newton's mechanics.

Whether we decide to set up a physics experiment in Peter's frame of reference on the platform, or Mary's on the train, is unimportant. We will get the same results. Indeed, we can readily transform our equations of motion from Peter's to Mary's frames of reference (or vice versa) by what is called a *Galilean Transformation*.

So, what is all the fuss about? It is surely just "common sense." Having just had our notion of common sense trashed in the previous Chapter, we should by now be inured to what is to come. To lead us in gently, let us first discuss two important results that were disturbing physicists, as the Nineteenth Century, with appropriate fireworks and popping of corks, passed over into the Twentieth.

6.4   Harbingers of a Scientific Revolution

An air of complacency pervaded physics at this time. Classical Mechanics, as formulated by Newton and extended by Lagrange and Hamilton, accounted successfully for terrestrial and celestial phenomena involving moving bodies and their collisions. Maxwell had completed a seminal work unifying electricity and magnetism into the single field of *Electromagnetism*. Some physicists believed that all important work in physics had been completed, and all that remained was to fill in the details.

However, a close inspection revealed that Maxwell's equations of electromagnetism are *not invariant under a Galilean Transformation*. This may sound erudite, but it is just physicists' jargon, which we shall now explain. The Galilean Transformation, which we encountered in the last Section, is the method we use to transform from equations in one frame of reference (e.g. Peter's on the platform) to those in another (e.g. Mary's on the moving train). What was discovered was that the predictions of Maxwell's theory depend on the frame of reference used, something that we have maintained is unphysical.

In other words, an experiment on electromagnetism would be able to reveal our *absolute velocity*, contrary to what was affirmed in the previous Section. So, who is right: Galileo or Maxwell? Or are they both wrong?

In the last Chapter, we saw that electromagnetic radiation can propagate in a vacuum. This is in contradistinction to sound, which requires a material such as air to maintain the wave motion, and sea waves, which of course require water. In the 19th Century it was thought that an as yet undiscovered, but all pervading, medium must exist to enable the propagation of light waves. This was called the quintessence[3] or ether (aka *aether*). According to the physics of Newton and Galileo, the velocity of light on earth with respect to the ether should be given by the same Galilean equations discussed in the previous Section for the motion of a person on a train (i.e. the velocity of light as measured on earth should be equal to the velocity of light through the ether plus or minus the speed of the earth along its direction of motion).

As we mentioned earlier, we are always in motion with respect to the ether, so the time taken for a ray of light to travel a certain distance in our direction of motion, and return, is expected to be different from the time taken to travel the same distance at right angles to our direction of motion, and return. It is analogous to a person swimming in a moving stream of water. The time taken to swim 100 m upstream and then return downstream to the starting point is different from the time taken to swim 100 m across the stream, and return. The huge magnitude of the velocity of light (300,000 kms per second) compared with our planetary speeds means that this time difference is very small, but we should still be able to measure it, if Newtonian physics is correct.

In a series of brilliantly conceived experiments, beginning in 1881, Albert A. Michelson and Edward W. Morley split a beam of light into two rays, sent these off on perpendicular paths, bounced them back from mirrors and then let the returning two rays produce an interference pattern on a screen. By this approach they were able to show that the velocity of light in the two perpendicular directions was actually *the same*. This did not seem possible. It was analogous to Mary, when measuring the velocity of the ball thrown by Peter in the last Section, obtaining a velocity of 20 kph, even though she was located on a moving train. It made no sense.

In 1892, the Dutch physicist, Hendrick Lorentz, remarked that the negative results of the Michelson-Morley experiments could be accounted for by replacing the Galilean Transformations by another, more complicated set of

---

[3]The fifth classical element after earth, fire, water, and air. The name has made a reappearance as a form of Dark Energy (see Chap. 11).

transformations, now known as the *Lorentz Transformations*. This explanation was, however, totally *ad hoc*, and as such only acceptable as a working proposition. In this regard, it was similar to Max Planck's *ad hoc* quantum explanation of black body radiation that we discussed in the last Chapter. The physics that underlay the Lorentz Transformations was supplied by the same young genius who had sorted out what lay behind Planck's mysterious quanta, in between carrying out his duties as a patent clerk in Berne, Switzerland.

6.5   Here Comes Einstein

In the *annus mirabilis* of 1905, the 26 year-old Albert Einstein published six papers, three of which would revolutionise the field of Physics. One of them [2] reintroduced the Lorentz Transformations as a consequence of a very daring assumption, i.e. *that the speed of light c had to be exactly the same for all inertial frameworks*, thus becoming a universal constant and rendering the ether superfluous.

So, why does all the credit go to Einstein and not to Lorentz?

Fame is a Fickle Food
Upon a shifting plate [3].

Not all the kudos in science finishes up where it rightly belongs, and a discussion of the pros and cons of the many examples would fill a lengthy monograph. However, in this case the reason for Einstein's fame is because his conjecture provides a full justification for the Lorentz transformations and leads us, as we shall see, to some astonishing conclusions.

What does it mean when we say that the speed of light is constant in all inertial frames of reference? Let us imagine that, in our example from the previous Sections, Peter, instead of throwing a ball towards Mary, directs a ray of light past her. He measures the speed of the light as it leaves him and obtains a value of c kph. Mary then measures the speed of the light as it passes her, and, because the train is moving at 10 kph, using Galilean transformations we would expect her to obtain a value of c-10 kph. However, she doesn't. She also obtains a value of c kph.

We would surely expect something of this kind if the speed of light were infinite. Adding or subtracting finite numbers from infinity yields infinity.[4]

---

[4]Mathematicians may cringe, but in Chap. 8 we shall learn that physicists have even resorted on occasions to subtracting infinity from infinity and obtaining a finite number.

However, although very large, c is not infinite.[5] The error of 10 kph, compared with the value of c, amounts to roughly 1 part in 100 million, and in most circumstances it could be completely disregarded. However, when, instead of balls, subatomic particles are shot out in nuclear reactions, they may be moving at speeds comparable with c. In this case the error obtained by using Galilean transformations would be quite large.

Indeed the consequences of *Special Relativity*, as the new field springing out of Einstein's conjecture was called by Einstein himself, are astonishing and manifold, resulting in some very bizarre effects, rivalling those we encountered with QM in the last Chapter. They made Einstein an instant (and sometimes controversial) celebrity, even among the general public. In the next Sections, we shall discuss briefly some of these effects: notably *time dilation, Lorentz contraction, the puzzle of simultaneity* and *mass/energy equivalence*.

6.6   Time Dilation

In the real world, as distinct from the world of fiction and fairy tales, time has always flowed inexorably from the past to the present, and at the same rate for everybody, no matter where they are or what is their state of motion. Or so it was believed, Giang Huong and Tu Thuc notwithstanding. However, the Special Theory of Relativity put an end to this complacency.

Let us remove Mary from the comfort of her railway carriage, train her as an astronaut, and place her on a high-velocity rocket ship, which can attain speeds comparable to (but never equalling or exceeding) that of light. When Peter looks into the spaceship moving away from him, he notices an amazing thing: everything and everyone on board appears to be in a state of slow motion. Even the cabin clock has slowed down. Comparing it with his own earthbound timepiece, Peter observes the second hand of Mary's clock completing only one revolution of the dial in the time it takes his own clock to complete two. Time is actually moving more slowly in Mary's frame of reference than in his own. She is indeed in a sort of Land of Bliss. This effect, which is known as *Relativistic Time Dilation*, is a direct consequence of the Lorentz transformations. However, even without using these transformations, the effect is easily deduced from Einstein's assumption that the speed of light is the same for all observers in inertial frameworks.

Imagine Peter is peering into Mary's spaceship, which is at rest on its launch pad. He observes a ray of light passing between two points in the ship

---

[5]To be precise its value is 299,792,458 m per second. This value is exact, as since 1983, the metre has been defined as the distance travelled by light in a vacuum in 1/299,792,458 s.

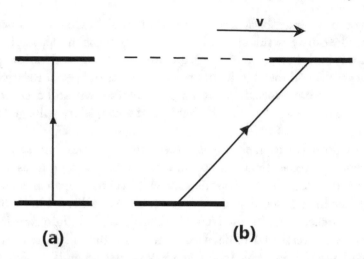

Fig. 6.2 Ray of light passing between two points in spaceship: **a** when ship is stationary; and **b** when ship is moving to the right with velocity v

(see Fig. 6.2a), and measures the time this takes. Later, after take-off, Mary turns the ship and flies by him at a velocity v. Peter again measures how long it takes for a ray of light to pass between the same two points in the ship (see Fig. 6.2b). In this case, however, because the spaceship is moving, the light ray must travel a larger distance before it reaches its target. Since the speed of light is the same in both cases, Peter concludes that in the second case the light takes longer to complete its journey. To Peter, time has effectively slowed down in the moving spaceship.

Now what does Mary think when she peers out of the spaceship rear window and sees Peter vanishing off into the distance? To her, everything in Peter's frame of reference is moving more slowly than in hers. She sees the second hand on his clock crawling around the dial, and suddenly realises that time itself is changing more slowly in Peter's surroundings than in her own. When she returns from her voyage, Peter will still be a young man, and she will be older.

Of course, a paradox is immediately evident: both Peter and Mary see the other person ageing less rapidly than themselves. They can't both be right. This contradiction is called the *twin paradox*, as it is usually formulated in terms of two identical twins travelling apart on separate space ships.

In mathematics and physics, a paradox is normally an indicator that we have made a mistake in our logic. Is Special Relativity wrong? No, but we have indeed violated the conditions under which it applies. Special Relativity deals with *inertial frames of reference* only, which means that Peter and Mary must be moving apart at a constant velocity for its provisions to apply. In that case, they can never get back together to compare their clocks or lament their wrinkles. The only way this can happen is if Mary turns the spaceship around and brings it back home. In doing so, she undergoes an acceleration, positive and/or negative, and thereby violates the conditions of Special Relativity. The paradox is therefore resolved. However, time dilation will still occur, and as a consequence of the acceleration she has experienced, Mary will age less than Peter.

Time dilation is extremely small, except for velocities v close to c. The energy required to accelerate a rocket ship increases rapidly as its speed approaches that of light, for reasons that will become apparent later in this Chapter. The highest speed attained by a rocket was achieved in 2006, when an Atlas V Rocket, carrying the New Horizons probe to Pluto, was launched from Cape Canaveral in Florida, USA. Its maximum speed, relative to the launch pad was 58,536 kph (about 16.3 kms per second), which yields a v/c ratio of 0.0000542. Accordingly, its relativistic time dilation effect was of the order of one in a billion, which is very small, but significant over a long journey.

The first direct test of time dilation for normal objects and velocities was carried out by Hafele and Keating [4] in 1971. They flew four caesium atom-beam clocks around the world in commercial jet airliners, both eastwards and westwards, and compared the clocks afterwards with reference clocks that had remained behind in the U.S. Naval Observatory. The clocks lost 59 ± 10 ns during the eastward flight and gained 273 ± 7 ns during the westward flight, compared with the land-based clocks. The difference between the eastward and westward results is explained by taking into account the rotation of the earth. These figures agree well with the predictions of Einstein's theory of Relativity.

It should be noted that the altitude of the airliner varied throughout the flight, but was generally around 9000 m. At this altitude, the gravitational force experienced by the clocks in the aircraft was less than that experienced by the ground-based clock. As we shall see in the next Chapter, gravitational fields produce another form of time dilation, which must be included in the analysis of the results of the Hafele-Keating experiment.

A routine application of Einstein's theories may be found in a device that most of us use regularly. Because of the difference in altitude and speed between GPS satellite transmitters and the receivers in the navigators of our vehicles and mobile phones, corrections for time dilation—due to the relative motion of the satellites and our phones, and to the different gravitational fields that they experience—must be included in the software that is used to calculate GPS positions.

When physicists explore the domain of cosmic rays, they encounter particles that are moving at speeds approaching that of light, when time dilation effects are much more pronounced. The muon is such a particle. In Chap. 9 we shall discuss muons, and other subatomic particles, in some detail. Here we need only note that the muon's average lifetime, when it is at rest, is 2.2 millionths of a second. In this time, if we disregard the time dilation of Relativity, even if the muon were travelling at the velocity of light, it would only travel 660 m. However, time dilation cannot be disregarded: thanks to it, physics students in their laboratories today routinely observe the arrivals of thousands of muons from the upper atmosphere, i.e. from distances of 15 km to more than 100 km.[6]

6.7   Lorentz Contraction

Suppose now that we manage to hitch a ride on a muon on its way from the upper atmosphere to the earth's surface below. We have done the calculations and are aware that with the muon's average lifetime of 2.2 µs, we will only have travelled an average of about 660 m before the muon decays beneath us. And yet when we were back on earth, we had observed most of the muons arriving safely, having travelled much further. It might seem that the outcome depends on which frame of reference we choose for our calculations. Doesn't this violate Einstein's edict that the results of experiments should be independent of the frame of reference?

No, because the Lorentz transformations contain another surprise: not only does the rate of passage of time vary with the speed, but so also does the length. An observer travelling at a certain velocity relative to an object will notice a decrease in the length of the object in the direction of motion.

There is a reciprocity between time dilation, and what is now called the *Lorentz contraction* of distance. Observer A in the laboratory on earth observes a long-lived (due to time dilation) muon travelling several kilometres to the earth; Observer B on the muon sees a short-lived muon travelling a much shorter distance (due to Lorentz contraction) to the earth's surface. Thus both observers see the muon arriving safely at the laboratory before the muon decays.

These concepts have now been taught in physics classes for the best part of a century. However, when first published by Einstein, they puzzled not only lay people, but specialists as well. Shortly after the publication of Einstein's

---

[6]In doing so, they are replicating a version of an experiment first performed by Bruno Rossi and David B. Hall in 1941.

paper, it was asserted by Croatian mathematician and theoretical physicist, Vladimir Varicak, trying to come to grips with the new theory, that the length contraction is real according to Lorentz, while it is only apparent, or subjective, according to Einstein. Einstein "clarified" the situation, as follows:

"The author unjustifiably states a difference between Lorentz's view and mine concerning the physical facts. The question as to whether length contraction really exists or not is misleading. It doesn't "really" exist, in so far as it doesn't exist for a co-moving observer, though it "really" exists, i.e. in such a way that it could be demonstrated in principle by physical means by a non co-moving observer" [5].

6.8   Simultaneity

Let us use our newfound knowledge of the principles of Relativity to carry out a thought experiment, along the lines that we employed in the last Chapter when we discussed the strange case of Schrödinger's cat. We consider a situation, as depicted in Fig. 6.3. A railway carriage, containing a youthful Mary, who has just returned from her space travels, is driven at rapid speed past an ageing Peter, watching from a platform. Mary is standing in the middle of the carriage, equidistant from both of its ends. At the precise moment that she passes Peter, she fires a device emitting a flash of light in both directions.

What follows from her point of view is straightforward: light from the flash travels at speed c, as measured by her, and reaches both ends of the carriage at exactly the same time. Nothing strange there. Now let us consider what Peter sees. The light travelling towards the rear of the train will reach the back end of the carriage *before* the light travelling forwards reaches the front

Fig. 6.3 Mary in the moving carriage fires a flash, the effects of which are also observed by Peter on the platform

end. Why? The reason is that the back end of the carriage is moving towards the flash, and thus the light has a shorter distance to travel, compared with the light travelling towards the front end, which is moving away. So, the two events (when the light impacts on each end of the carriage) appear to be simultaneous to Mary, but not to Peter, who sees light arrive at the rear end first.

The implications arising from the dependence of the simultaneity of events on the state of motion of the observer are profound. Will someone watching a murder from a speeding carriage see the victim die before the trigger of the revolver is pulled? Relativity protects us from such absurdities. Events that are *causally connected*, i.e. where one event is a direct consequence of the other, cannot have their order reversed by Lorentz transformations. All observers, irrespective of their motion, see these events in the correct sequence.

However, we saw in the last Chapter when discussing Entanglement, that Quantum Mechanics plays free and easy with the concept of simultaneity. When an observation specifies the state of one of a pair of entangled particles, the state of the other is simultaneously specified, no matter how far away that particle is located. There is a disagreement here with the Theory of Relativity, and it cannot be swept under the carpet. We shall discuss this point further in Chap. 12, Part 3.

Let us now move on to the final topic in our *brave new world*. It is so momentous, not just to physics, but to the history and future of humankind, that we break the rule we have set ourselves for this book: we allow ourselves just one mathematical formula.

6.9 The Only Formula in This Book

In fact, it is such an important formula, and it looks so *elementary* that it can be found everywhere, even on the T-shirts of people who have no idea of what it means. We refer, of course, to:

$$E = mc^2,$$

where E stands for energy, m for mass and c is, as usual, the velocity of light. The formula states the equivalence between mass and energy, with a coefficient c^2 in the transformation from the latter to the former.

Before Einstein, there existed two separate conservation laws in chemistry and physics; i.e. the *Law of Conservation of Mass*, and the *Law of Conservation of Energy*. In chemical reactions, the total mass of the reactants had to be equal to the total mass of the reaction products. In physics, it had long been known that energy can be converted from one form to another (e.g. *potential energy* to *kinetic energy* to *heat*), but never created or destroyed. Since Einstein, these two quantities, mass and energy, can be considered as different forms of a single entity. In nuclear reactions, mass can be converted to energy, and vice versa. The same is also true in chemical reactions, but the energies involved are so small that any change in mass of the reactants is unobservable.

Einstein's formula is remarkable from a philosophical point of view because it unexpectedly relates together three variables which nobody had ever before dared to assume were associated, i.e. mass, energy and the velocity of light. It may be interesting to some readers to recall that a similar kind of unexpected association between apparently unconnected quantities can be found in mathematics. We refer to Euler's formula, which establishes a fundamental relationship between the hitherto unrelated fields of trigonometric functions, exponentials and complex numbers.[7] The physicist, Richard Feynman, called Euler's equation *"our jewel"* and *"the most remarkable formula in mathematics"*. It may be a subject of philosophical speculation to investigate the underlying reasons for these (and a few other) unexpected connections.

A formal demonstration of the mass/energy relationship is beyond the scope of our book, since it involves a lengthy mathematical treatment. (A somewhat simpler justification of it can be found in [6].) What is more instructive for our purposes here is to examine some of its consequences.

We begin by considering the problem of an accelerating rocket ship. In classical physics, as the rocket consumes fuel and ejects the combustion products, it receives a propulsive force which drives it forwards. As long as this force remains, the rocket will continue increasing its velocity, and thereby its kinetic energy. According to Einstein, however, some of this kinetic energy will be transformed into mass, thus making the rocket ship more difficult to accelerate further. The closer the rocket's velocity gets to the velocity of light, the more pronounced this effect becomes. If the rocket's speed were ever to reach c, its mass would be infinite. This is mathematics' way of telling us that it is not possible ever to accelerate a rocket, or any other material object, to speeds greater than or equal to c. The velocity of light is a universal limiting speed that nothing can exceed.

One may raise the objection that photons are particles, and surely, since they are particles of *light,* they travel at speed c. Yes, they do, but they are a special type of particle, possessing no *rest mass.* The rest mass of a particle is, as the name implies, the mass of the particle when it is stationary. When it is in

---

[7]Those of a mathematical bent may check this out in any elementary mathematics textbook, where they will find: $e^{ix} = \cos x + i \sin x$.

motion, its total mass is greater than its rest mass because of the contribution from the conversion of kinetic energy into mass, as we have discussed in the previous paragraph. Photons, having no rest mass, must spend their entire existence zipping hither and thither at speed c. They can never slow down.

This upper limit on the speed at which material objects, e.g. astronauts, can travel has long been the bane of science fiction writers, as it puts restrictions on how far one can travel in a human lifetime. Given that the closest stars, located in the Alpha Centauri system, require 4.3 years of travel at speed c to reach them, it is clear that by far the vast majority of the cosmos is beyond the reach of humankind. The invention of Hyperspace, a region outside the realm of Relativity, has become an accepted method in the domain of science fiction to overcome the limits imposed by Einstein's theory. In the real world, we can only sit and peer through our telescopes in frustration or find something else nearer to home to sate our curiosity.

Let us now put our wistfulness aside and return to the application of Einstein's formula to more earthly pursuits. Its importance and huge impact on the history of the twentieth century is due to the magnitude of the coefficient c^2 in our lonely equation. To give a simple example, if one kilogram of matter collides with one kilogram of antimatter (we will discuss antimatter in Chaps. 8 and 9), they annihilate each other: the two masses disappear and as much as fifty billion kilowatt-hours (kwh) of energy are produced.

To put this number into perspective, we recall that in the oil industry, the unit *toe* (tonne of oil equivalent) is defined as the amount of energy released by burning one tonne of crude oil, and is approximately equal to 11,600 kwh. Hence the annihilation of 1 kg of matter and 1 kg of antimatter would yield the same energy as burning 4.3 million tonnes of crude: i.e. almost enough to provide for the entire energy needs of the United States for one day (about 6 million toes).

In reality, however, kilograms of antimatter are difficult to find, and the practical transformation of mass into energy relies on one of the two processes of nuclear fission and nuclear fusion. The result is either nuclear energy, released under controlled conditions and deployable for practical purposes, or a violent burst of a huge quantity of destructive energy. The latter is the atomic bomb, if nuclear fission is employed, or the H-bomb, with the potential for an even more devastating effect, if nuclear fusion is adopted.

Given the consequences, should we blame Einstein (or science) for having incautiously opened such a *Pandora's Box*? Or rather humans, for their stockpiling of a terrifying arsenal of weapons of mass destruction? This question has been debated for the past seventy years, and we shall not discuss it further here.

References

1. For origin of limerick, see Appendix 6.1
2. Einstein A (1905) Zur Elektrodynamic bewegter Körper. Ann Phys 17:891–921
3. Dickinson E, 1830–1886
4. Hafele JC, Keating RE (1972) Around-the-world atomic clocks: predicted relativistic time gains. Science 177:166
5. Einstein A (1911) Zum Ehrenfestschen Paradoxon. Eine Bemerkung zu V. Variĉaks Aufsatz. Physikalische Zeitschrift v12: 509–510
6. Barrett RF, Delsanto PP, Tartaglia A (2016) Physics: the ultimate adventure (Chap. 7.4) Springer, Berlin

7

General Relativity

7.1 Let's Try It This Way[1]

Sarah woke up with a headache, a foul temper, and an empty bottle of chardonnay on the bedside table. A loud snore emanated from the pile of bedclothes next to her.

The faculty colloquia didn't usually end this way, but Jim had brought home a group of friends last night to celebrate his promotion. Associate Professorship, no less. The deal had been clinched by his presentation: *"Use of the Future Subjunctive in Pre-Chaucerian Couplets, circa 1333."* Standing applause. All that remained were a few formalities to be settled this morning.

She squinted at her out-of-focus watch. Ten past nine. Suddenly she was wide awake. Jim's appointment was for ten-thirty. She grabbed his shoulder, and shook him. "Wake up!" she shouted. "Come on. Wake up, you drunken slob!".

"Huh!".

"Ten-thirty. Your appointment."

"So what?".

"It's ten past nine already."

Now she had his attention. He sat bolt upright, and peered at his own watch. Six-twelve. He reached his wrist across so that

---

[1] According to Kurt Vonnegut, the last words spoken by the human race will be when one scientist turns to his colleague and says, "Let's try it this way…".

© The Author(s), under exclusive license
to Springer Nature Switzerland AG 2021
R. Barrett and P. P. Delsanto, *Don't Be Afraid of Physics*,
https://doi.org/10.1007/978-3-030-63409-4_7

she could read the time. "Well obviously it's stopped. Possibly something to do with you dropping it into Angela's martini."

They were both awake now. Jim got out of bed, and spooned instant coffee into two mugs. All this would change when his promotion came through. Then it would be an espresso machine and freshly roasted beans from Kenya. He switched on the T.V. to check the time. The screen was full of snow, and the hiss of white noise emanated from the speakers. Now this *was* unusual.

A voice broke through the crackle: "… all the reports coming in have the same basic theme. People went to bed in one place and woke up in another. One man claims that he woke up next to his best friend's wife."

Jim chuckled, thinking of Angela. His smirk vanished as Sarah joined him.

"What's wrong with the T.V.?" she asked.

"Shh! Listen."

"The nation's transport system is in chaos. … One moment, please. We're getting Professor Karl Profundis on the line."

"Ach!" Jim scoffed. "He's the science bozo from our college. Nutty as the proverbial."

"Professor, you have an explanation for what's going on?".

"This is exactly what I said might happen. Read my book. Space–time is the fabric of the universe. In the neighbourhood of a black hole, it becomes stressed and warped. It's starting to break down, just like any other material. Wormholes are a distinct possibility."

Sarah pulled back the curtains and peered through the grimy window. Jim joined her, and together they stared at the scene outside.

Enormous holes were opening up everywhere in the street, and buildings were collapsing into them. An unearthly blackness existed in some directions, containing not even a flicker of light. In the sky, scores of people were floating around on a mad carousel, clinging as hard as they could to what was surely a fragment of space–time.

The carousel dipped and veered towards them. As it swung around Jim noticed a group of revellers, oblivious to what was happening, sloshing drinks and shouting to each other across a coffee table laden with unappetising finger food. They looked familiar.

Sarah turned to him. "But there are no black holes near the earth. ... Surely."

Jim shrugged. It wasn't his field. "Not unless someone decided to make one."

In the next instant, Jim's world vanished. He felt himself falling into a black silence, so intense that it swallowed even the beating of his heart, where seconds and centuries, metres and kilometres swirled together in a cosmic potpourri.

* * *

We defer our discussion of black holes until later in this Chapter (Sect. 7.5). By then the reader will be in a better position to decide whether the story of Jim and Sarah is fantasy or prophecy.

7.2 A 20th Century Cathedral

As we have seen in the previous Chapter, in 1905 Einstein proposed new concepts of space and time that overturned the credos of centuries. Two hitherto different quantities, space and time, were combined into the single entity: *space–time*. Profound physical consequences were found to arise from one basic hypothesis: that the speed of light in a vacuum is the same for all non-accelerating observers.

Immediately, the restrictive nature of this hypothesis raises a question: what happens when the observer *is* in an accelerating frame of reference? We have intimated, when discussing the twin paradox in the last Chapter, that consequences do arise, which are not at all obvious. The other outstanding issue, untreated in Special Relativity, was gravity, probably the most ubiquitous source of acceleration throughout the cosmos.

It was clear to Einstein that Newton's theory of gravity[2] was incompatible with Special Relativity, and in need of modification or replacement. Newton's equations contain no reference to time: the gravitational attraction between two objects depends only on their mass and their separation. According to Newton, a movement of one mass results in an instantaneous change in the force being felt by the other. This perturbation in the force propagates at an infinite velocity, which violates the limit of c imposed by Special Relativity. Einstein worked on these problems for a decade, and finally in 1916

---

[2]When Newton propounded his theory of Gravity to the Royal Society in 1686, Robert Hooke, a contemporary of his and a gifted scientist in many different fields, immediately claimed that Newton had plagiarised the law from him. The controversy continues to this day.

published the results of his deliberations in a paper that has become a classic in the literature of physics.

While Special Relativity is very important, both for its scientific relevance and for its applications, General Relativity owes its significance mostly to its philosophical value. In fact there were very few practical applications of General Relativity in the first fifty years of its existence, at which time, as we shall see, solutions of Einstein's equations were found for black holes, and other celestial objects. Prior to this period, the vanguard of physics lay with Quantum Mechanics, the other great 20th Century revolution. Purely as an intellectual monument to human ingenuity, General Relativity can be compared to other momentous achievements of humankind, such as the ninth symphony of Beethoven, Leonardo's *Giaconda* and the Egyptian Pyramids.

7.3   The Principle of Equivalence

Let us begin our intellectual journey into General Relativity, as Einstein did, with another thought experiment. Consider a passenger travelling in an elevator, as shown in Fig. 7.1.

If the speed of ascension in the first example in Fig. 7.1 is constant, then the passenger is in an inertial frame of reference, and the theory of Special Relativity applies, as we saw in the last Chapter. However, suppose that

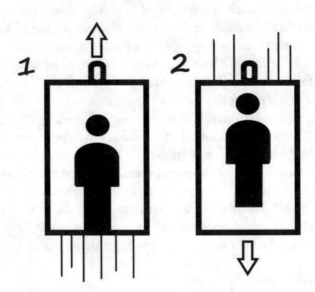

Fig. 7.1 Passenger in rising elevator (1) and free falling elevator (2)

the elevator motor is applying a constant acceleration to the elevator. The passenger will feel in his legs that he is becoming heavier. It is as though the force of gravity has increased.

Now imagine that the elevator is allowed to fall freely, as in the second example in Fig. 7.1. The passenger will drift about inside the elevator, as if gravity has suddenly been turned off. It hasn't, of course; it is just that gravity accelerates the cabin and the passenger downwards by equal amounts. Another way of looking at this is to say that the elevator and passenger are both falling at the same rate. This is what is happening in those videos we have all seen of astronauts floating and tumbling around inside satellite space laboratories.

These ideas owe their origin to Galileo, who noted that, ignoring air resistance, different objects dropped from a height at the same time would hit the ground together, irrespective of their mass. He realised that mass appears in two different guises: *gravitational mass*, which is responsible for an object's weight, and *inertial mass*, which is a measure of its resistance to changes in its state of motion. If you weigh an object, you are measuring its gravitational mass; if you push an object to set it in motion, you are resisted by the object's inertial mass. The fact that objects fall at the same rate, irrespective of their gravitational mass, told Galileo that their gravitational masses and inertial masses have the same value. This property is known as the *Principle of Equivalence.*

The genius of Einstein lay in his ability to pursue the consequences of his hypotheses, wherever they happened to lead him, with no regard to conventional thinking or *common sense*. His tool was rigorous logic, of the type that we have discussed in Part 1 of this book. In formulating Special Relativity, Einstein had asserted that there was no experiment observers could perform to determine whether they were in motion or not. He now maintained that there was no experiment observers could perform to determine whether they were in an accelerating elevator or under the influence of gravitational fields. This hypothesis became the basis of his theory of General Relativity.

However, in this case we must attach a caveat. Einstein is assuming here that the elevator and its immediate surrounds are small compared with the distance to the centre of the external gravitational field. If our elevator had a size of many kilometres and were located near the surface of the earth, then clearly the gravitational field experienced by the observer would be greater at the bottom of the elevator than at the top. This difference could be detected by the observer, as it results in a stretching force, or *tidal force*, acting on the observer.

With a small elevator, the difference in the gravitational field between the top and bottom of a normal elevator is too small to be measurable, and Einstein's hypothesis holds. It is analogous to the assertion that a football field is flat, even though we know it to be a part of the earth's curved surface. The curvature over the length of the field is too small to measure easily.

7.4   Bending of Rays of Light Under Gravity

As an example of the application of Einstein's hypothesis, let us now explore the question, raised by Newton, of whether light is deflected by a gravitational field: *"Do not Bodies act upon Light at a distance, and by their action bend its Rays, and is not this action … strongest at the least distance? [1]"* We know that, when we throw a ball, it falls to the ground, no matter how hard we try to throw it. What about the beam of light from our flashlight? Does it fall to the ground too?

Let us imagine that a ray of light is fired horizontally in a rising elevator, as shown in Fig. 7.2. First we assume that our elevator is in outer space, where there is no gravitational field, and by "horizontal", we mean perpendicular to the direction of motion of the elevator.

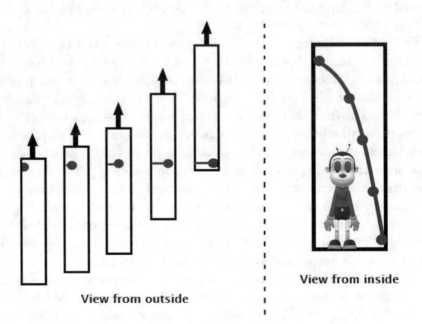

View from outside

View from inside

Fig. 7.2 Passage of light ray across an accelerating elevator

An external observer will see the light traversing from left to right, as the elevator accelerates upwards. This is shown in the left half of Fig. 7.2, where each drawing shows the elevator at equally spaced intervals of time. However, a passenger in the cabin will see the ray falling down, following a parabola, as shown in the right half of Fig. 7.2.

This all seems fairly straightforward. However, suppose now that we remove the accelerating drive from the elevator, and replace it by an equivalent downwards gravitational force. An external observer will see that the elevator is now no longer accelerating, but what about the elevator's passenger? According to Einstein, *since there is no difference between the effects of acceleration and gravity*, the ray of light will travel on the same parabolic path as before.

Of course, because of the enormous speed of light compared with the velocity of the elevator, the distance that the ray is deflected by the gravitational field in this instance would be infinitesimally small. However, in astronomical examples—e.g., a ray of light from a distant star passing close by the sun—the effect should be observable. This argument was seized upon as a possible check on the validity of General Relativity.

Before discussing astronomical observations, let us recall what Newton's law of gravity states on the same subject. As we saw in the last Chapter, the debate about the particle versus wave theory of light had been going on for centuries when Einstein began working on General Relativity. Maxwell's equations (see Chap. 5) represent the zenith of the wave theory's popularity, before the advent of Quantum Mechanics. There is no place for gravity in Maxwell's theory, a failing it shares with QM. There is therefore no mechanism within it for calculating any deflection of a ray of light by a massive object.

However, Newton was a believer in the corpuscular theory of light: "*Are not the Rays of Light very small Bodies emitted from shining Substances?*" [2]. The corpuscles of light (now called photons) have zero rest mass. As, since Galileo, it has been known that the path of objects in a gravitational field is independent of their mass, we would expect all particles, even those with zero mass, to fall along the same trajectory.

Indeed, Einstein in his first calculation of this effect in 1911 followed precisely this argument. However, by the time of his 1916 paper, he had realised that he needed to include the effect of space–time curvature (see Sect. 7.4), induced by the presence of the gravitational mass, and corrected his earlier calculation [2] The amended value for the deflection was twice as large as that of his first (and Newtonian) result.

Fortunately for Einstein, three years after his paper appeared in *Annalen der Physik*, Nature cooperated with him by providing on 29th May, 1919 one of the longest total solar eclipses of the 20th Century. A British astronomer, Sir Arthur Eddington, one of the few scientists outside the German-speaking world familiar with Einstein's work, led expeditions to the tropics of Brazil and to the island of Principe, off the west coast of Africa, to observe the eclipse, an image of which is shown in Fig. 7.3. The stars of the Hyades cluster were near the solar limb at this time, and provided a perfect opportunity to test the theory of General Relativity.

Eddington's analysis of the results confirmed that the stars were displaced from their usual positions by the amount predicted by Einstein. When the New York Times published the news, Einstein's reputation soared, and he became an international celebrity. A headline from the *New York Times*, November 10, 1919 is shown in Fig. 7.4.

One should not gain the impression that the bending of light predicted by General Relativity is something anyone can observe in their backyard with a pair of binoculars. The deflection of the stars of the Hyades cluster was predicted to be 1.75 arc seconds. This is equivalent to the angle subtended by

Fig. 7.3 Solar eclipse, 29th May 1919, observed by Sir Arthur Eddington, and providing the first evidence for Einstein's Theory of General Relativity. The horizontal bars to the top-right of the image indicate the position of stars. Image, public domain due to age (https://commons.wikimedia.org/wiki/File:1919_eclipse_positive.jpg (accessed 2020/05/26))

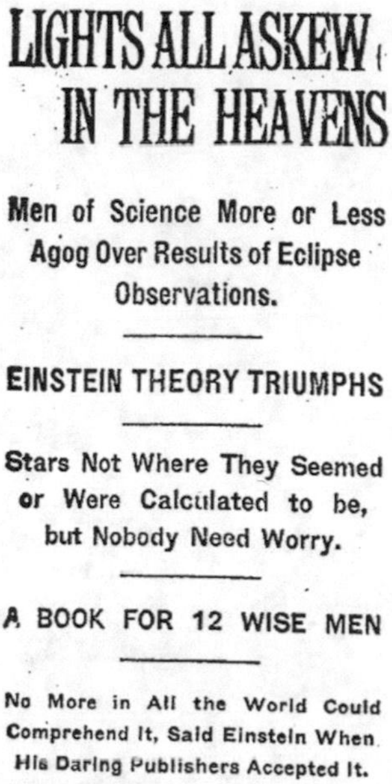

LIGHTS ALL ASKEW
IN THE HEAVENS

**Men of Science More or Less
Agog Over Results of Eclipse
Observations.**

EINSTEIN THEORY TRIUMPHS

**Stars Not Where They Seemed
or Were Calculated to be,
but Nobody Need Worry.**

A BOOK FOR 12 WISE MEN

No More in All the World Could
Comprehend It, Said Einstein When
His Daring Publishers Accepted It.

Fig. 7.4 Headline from the New York Times, November 10, 1919, announcing the results of Eddington's expedition. Image: public domain due to age (https://commons.wikimedia.org/wiki/File:Einstein_theory_triumphs.png (accessed 2020/05/26))

a 2 cm diameter coin placed at a distance of 2.4 km. Eddington's observations were 1.61 ± 0.30 arc seconds, which is within observational error of the predicted value. A controversy (since resolved) developed over Eddington's data analysis. For a more accurate analysis of Eddington's observations, see [3].

Confirmation of Eddington's results was sought by a second expedition in 1922 to observe the solar eclipse on September 21. A proposed observation at Wallal, in northwest Australia, was vigorously supported by Alexander Ross of the University of Western Australia, but met strong opposition in Britain. Ross pointed out that it had only rained at Wallal in September twice in the previous 25 years, surely a relevant factor in the choice of a location for such an important observation. British and Dutch-German expeditions, however, duly proceeded to a Christmas Island observation site, where it clouded over on the day, preventing any eclipse photography. Ross took part in a US Lick Observatory expedition to Wallal, successfully obtaining measurements for over 100 stars, and providing further confirmation of Einstein's theory.

A feeling for the post-World War I ambience of research into General Relativity can be obtained from Figs. 7.5 and 7.6 below, and comparing them with Fig. 7.11, a modern observatory for the detection of gravitational waves.

Fig. 7.5 Donkey team hauling equipment from a schooner during the 1922 Solar Eclipse Expedition at Wallal. Image: courtesy of State Library of Western Australia (Image: Sourced from the collections of the State Library of Western Australia and reproduced with the permission of the Library Board of Western Australia. https:// purl.slwa.wa.gov.au/slwa_b4792891_1 (accessed 2020/05/26))

Figure 7.5 shows equipment being manhandled from a grounded schooner onto a cart drawn by a donkey team at Wallal during the 1922 solar eclipse expedition. Nine metre tides are the norm in this region of Australia. A rude twig shelter provides a modicum of shade from the tropical sun for a hanging canvas water bottle. Figure 7.6 shows Dr Robert Trumper, expedition member, observing the sun's image.

Eddington was a strong supporter of General Relativity. Shortly after the publication of Einstein's paper, he delivered a lecture presenting the new theory. At that time it was said that only three people in the world understood Einstein's work.[3] One of the audience asked him if he was one of these three.[4] Eddington stopped and seemed to puzzle a while, until urged by the questioner not to be modest. "On the contrary," he quipped. "I'm trying to think who the third person might be."

---

[3]This number is even smaller than the 12 wise men in the NY Times headline in Fig. 7.4.
[4]Allegedly related by Ludwik Silberstein, a Polish-American Physicist, 1872–1948.

Fig. 7.6 Observation of sun's image through 5-foot cameras at Wallal Downs Station. Image: courtesy of State Library of Western Australia (Image: Sourced from the collections of the State Library of Western Australia and reproduced with the permission of the Library Board of Western Australia. https://purl.slwa.wa.gov.au/slwa_b4833949_1 (accessed 2020/05/26))

7.5   Einstein's Field Equations

Following his recognition of the importance of the Principle of Equivalence, Einstein's next step in the development of General Relativity was his realisation that space–time, which he had introduced in his Special Theory, was not necessarily flat. By this we mean that the geometry of Euclid, that we are all familiar with from high school, need not apply.

His conclusion was that the presence of a large mass distorted space–time in the same way that a heavy person standing on a trampoline distorts the trampoline's surface. The effect is illustrated in Fig. 7.7. The result is that any object travelling past a heavy mass is deflected from what would otherwise have been its path.

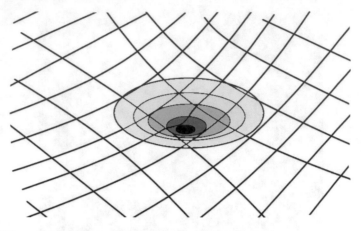

Fig. 7.7 Distortion of space–time by the presence of a heavy mass

In effect, what Einstein proposed was that the mysterious "action at a distance" of Newton's gravity was actually the result of a distortion of space–time, the fabric of the cosmos. He applied a branch of mathematics, called tensor analysis, to his formulation of General Relativity. The equations he developed, now known as Einstein's Field Equations, supplant the equations of Newton in the formulation of a theory of gravity.

Why then are Newton's equations still one of the first things that students of physics learn at the beginning of their courses? Firstly, they are very much simpler than Einstein's to apply. This would be of no consequence, if they gave the wrong results. However, as we have seen in Chap. 2, "wrong" to a physicist means something quite different from "wrong" to a mathematician. In terms of accuracy, for most applications, the results of the two theories are indistinguishable within the limits of observational error. Only when very high velocities are involved (approximating that of light) and/or very large masses are present, do we need to turn to Einstein.

The major area of current application of General Relativity is in the development of theories for the origin and evolution of the universe, as we will see in Chaps. 10 and 11. Here we present only a few special examples of simple minded solutions.[5] However, a qualitative description of some of the results is important for an understanding of modern cosmology. Obtaining these solutions generally requires some simplification of the problem, usually by choosing examples with particular symmetries that reduce the number of variables involved.

---

[5]The reader should not panic: as was well recognised by Einstein, the *Field Equations* are very difficult to solve analytically, and are well beyond our scope here.

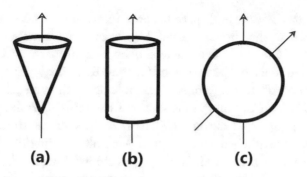

Fig. 7.8 Examples of three symmetrical 3D figures, with axes of symmetry shown

We are all familiar with symmetries in our everyday lives. Our faces are almost, but not quite, symmetrical about a vertical line drawn down through our noses. A cylinder is symmetrical with respect to rotation about an axis that passes through its middle; a sphere is symmetrical with respect to rotation about any axis passing through its centre. Examples of a few symmetries are shown in Fig. 7.8.

The advantage of symmetries in physics is that they reduce the complexity of the mathematics involved in the application of physical laws to the problem. For example, to specify the surface of a sphere, one needs to know only its radius and the location of its centre, while to specify a general surface in 3D space, one needs to know how the surface varies in all three spatial directions, i.e. a virtually infinite number of data.

The first solution to the Field Equations was provided by Karl Schwarzschild in 1916 for a spherically symmetrical mass with no electric charge and no angular momentum. Think of a heavy ball that is not spinning. This model provides a useful description of stars and planets. In the limiting case of small masses, the solution reverts to Newton's law, as it must if the Field Equations are correct.

However, Schwarzschild's solution reveals a singularity[6] at the origin (i.e. at the centre of the gravitational mass), if the matter comprising the sphere is dense enough. Singularities are common in mathematics, but they are causes of concern in physics. In this case, the singularity is interpreted as a *black hole*. At a distance, called the Schwarzschild Radius, from this central singularity lies a surface known as the *event horizon*. Any physical object at a distance less than the Schwarzschild Radius (i.e., within the event horizon) will collapse under the intense gravitational field and become captured by the black hole. Nothing, not even light, can escape from inside this region.

---

[6]A singularity is a point at which a function takes an infinite value.

The very natures of time and space become intermingled here, as in the plot of a far-fetched science fiction movie. However, as we shall see later in this Chapter, black holes have been detected by astronomers, and are now an accepted consequence of the theory of General Relativity.

The possibility of creating a mini black hole on earth in an accelerator, such as the Large Hadron Collider in Geneva, has been raised, with some concern that it might escape and devour the earth. At first it was thought that the accelerator had insufficient energy for black hole production. More recent work shows that such an event is a real possibility. Luckily the calculations indicate that such tiny black holes would pose no danger, since they would soon shrink and disappear before being able to gobble up any matter [4, 5].

The second solution of the Field Equations that we wish to consider here is the one for a homogeneous, isotropic universe. This is just physicists' jargon for a universe that is not "lumpy", and looks the same in all directions. One might well argue that our universe, comprising, as it does, stars, galaxies of stars and clusters of galaxies, is indeed "lumpy". From our perspective here on earth, it most certainly is. However, by moving far enough away and taking a big picture, these inhomogeneities merge into each other in the same way that the pixels on a TV screen merge together, unless one sits too close. They should not affect the results of the model *too* much.[7]

Solutions for this model of the universe were developed independently by Alexander Friedmann, Georges Lemaître, Howard P. Robertson and Arthur Geoffrey Walker in the 1920s and 1930s. The model is generally known, rather unimaginatively, as the FLRW model, and is discussed in detail later in Chap. 10. One of its characteristics is the presence of a singularity which is identified with the big bang at the origin of the universe.

A third important solution for the Field Equations was developed for *rotating* massive objects, including black holes, by New Zealand mathematician, Roy Kerr, in 1963. His work inspired a flurry of activity on the physics of black holes by Stephen Hawking, and others. One of the unusual effects arising from Kerr's solution is *frame-dragging*. Objects in the vicinity of a rotating black hole become caught up in the rotation because of the curvature of space–time that the rotation generates. At close enough distances, even light itself must rotate with the black hole.

As we have already remarked, exact solutions of the Field Equations are hard to find. However, even in classical physics, not all problems that appear on the surface to be simple, can be solved precisely. The three-body gravitational problem, where the earth, sun and moon interact with each other, is

---

[7] *Optimism is the faith that leads to achievement. Nothing can be done without hope and confidence:* Helen Keller.

a well-known example that we have discussed in earlier chapters. However, the enormous progress in computational technology since the latter half of the 20th Century has facilitated an alternative approach. Physicists now use numerical techniques and fast computers to obtain the trajectories of the planets and space probes with the desired accuracy. Likewise, numerical techniques are now being employed in *"numerical relativity"*, which is a burgeoning research field.

Being as it is, a theory of gravity, General Relativity is of most importance in domains where gravity overweighs the other forces of nature, i.e. in the cosmos at large. In this domain, experimentation, as it usually pertains to physics, is very difficult, and astronomers largely rely on observations of events over which they have no control. Nevertheless, by chance, natural phenomena do arise from time to time that provide an opportunity to subject the predictions of General Relativity to observational test. We shall discuss some examples of these in the following Sections.

7.6 Further Observational Evidence for General Relativity

Changes in Orbit of Mercury

Since the time of Johannes Kepler (d. 1630) it has been known that the orbits of the planets about the sun follow an elliptical path. Newton showed how his law of gravity explained this phenomenon. However, closer observation revealed that the major axis[8] of the orbit of mercury rotates slowly about the sun. The effect, much exaggerated, is shown below in Fig. 7.9.

The vast majority of this precession is accounted for by Newton's theory, and the effects of the outer planets (see Appendix 7.1). However, there remains a 40 arcseconds/century discrepancy, which is very well explained by General Relativity, when one includes the distortion of space–time caused by the gravitational mass of the sun.

The explanation of the hitherto mysterious anomaly in Mercury's orbit, combined with Eddington's observation of the bending of light by gravity, remained virtually the only experimental tests of General Relativity for the best part of half a century. In 1916 Einstein had proposed three possible tests of his theory [6]. The third, known as the gravitational redshift, was not carried out successfully until 1954.

---

[8]An ellipse has two axes of symmetry. The larger of these is known as the "major axis".

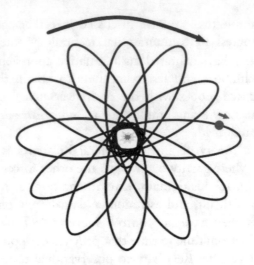

Fig. 7.9 Precession of the orbit of mercury about the sun (schematic)

Gravitational Redshift of Light

Consider the situation where we fire a projectile upwards against a strong gravitational field. As the projectile rises, it loses kinetic energy because it must do work against the gravitational force that is trying to pull it back to the ground. Now consider what happens if we replace the projectile by a series of photons (aka a beam of light) from a laser. We have already seen in Sect. 7.3 that photons are deflected by a gravitational field in the same manner as other particles. We might therefore expect them in this example to lose energy as they struggle to overcome the gravitational force retarding them.

There is, however, an important difference between photons and other projectiles: photons cannot lose energy by slowing down, as does a bullet fired vertically. Instead they are condemned always to travel at speed c. However, this does not imply that the Law of Conservation of Energy is somehow broken in this example. We have already seen in our Chapter on Quantum Mechanics that the energy of a photon is proportional to its frequency. When a beam of light travels from a region of strong gravitational field to a weaker one, the photons manage to achieve energy conservation by reducing their frequency, while still travelling at the velocity c. This phenomenon is known as the gravitational redshift. It is called a redshift because the light is shifted to lower (i.e. redder) frequencies.

The first reliable verification of the gravitational redshift was made in 1954 by Popper [7], who measured a frequency shift of 0.007 percent in light emitted from the white dwarf star, 40 Eridani B. His result was within experimental error of the value predicted by General Relativity. Since this first ground breaking observation, numerous more recent experiments have confirmed the existence of gravitational redshift (aka *gravitational time dilation*) (See Appendix 7.2).

Gravitational Lensing

In Fig. 7.10, we have an example, captured by the Hubble Telescope, of a very beautiful, but rare, phenomenon in astronomy known as an *Einstein Ring*.

The photograph shows a galaxy cluster containing hundreds of galaxies interacting with each other through their gravitational fields. The galaxies are the fuzzy objects distributed throughout the image. Some are clearly spiral, similar to our Milky Way. Others are seen as thin discs, and are probably spiral galaxies with their edges pointing towards us.

In the centre of the image are the graceful arcs of an Einstein Ring. This phenomenon is caused by the bending of light from a distant star or galaxy by the gravitational field within the cluster. As a consequence, the distant object will appear to lie in a different location from its true position.

Fig. 7.10 The galaxy cluster, SDSS J0146-0929. The graceful arcs near the centre are an example of an Einstein Ring (see text). Image, courtesy of ESA/Hubble & NASA (Acknowledgment: Judy Schmidt. https://www.nasa.gov/image-feature/goddard/2018/hubble-finds-an-einstein-ring (accessed 2020/05/27))

The effect is known as *gravitational lensing*, and is explained in more detail in Appendix 7.3.

7.7   Gravitational Waves

We have mentioned earlier that Newton's law of gravity contains no reference to time; it therefore implies that if a mass distribution somewhere out in space changes suddenly, the gravitational effect is felt instantaneously throughout the universe. According to Einstein's Theory of Relativity, however, the propagation of such effects cannot take place at a speed greater than that of light.

One of the consequences of General Relativity predicted by Einstein is that the effect of any sudden change in the distribution of mass is propagated through space in the form of a gravitational wave. This wave is a ripple in the fabric of space–time. In our analogy with a heavy mass on a trampoline in Fig. 7.7, imagine the mass being suddenly displaced up or down by a small amount. A wave induced by the movement in the elastic trampoline surface will propagate across the trampoline at a speed determined by the nature and tension of the material comprising the trampoline. In the case of gravity, the gravitational wave is expected to propagate across the space–time "fabric" at the speed of light. However, the magnitude of such space–time distortions was predicted to be so small as to be unobservable with the technology available in the first half of the 20th Century (i.e., during Einstein's lifetime.) During this period, gravitational waves were generally considered an intellectual curiosity of little relevance to mainstream physics.

This situation changed dramatically in the 1960s when Joseph Weber at the University of Maryland announced the detection of gravitational waves in a series of scientific papers. He used "antennas" made from aluminium bars two metres long and one metre in diameter. They were located at two sites separated by approximately 1000 km, and detections were triggered when perturbations to the bars were detected simultaneously at the two locations.

The initial excitement of the physics world soon became tempered by a feeling that all was not well. Weber's detections raised more questions than they answered. Theorists began examining what sort of event could produce gravitational waves of a magnitude sufficient to register on Weber's detectors. Calculations indicated that all the stars in the universe would need to fall into black holes to explain his detections. Over the next decade, other experimentalists designed and carried out more sensitive observations than Weber's but were unable to reproduce his results. His anomalous findings were attributed to over-enthusiasm and a lack of healthy scepticism.[9]

Although over the next forty years no success was achieved in detecting gravitational waves directly, indirect evidence of their presence soon began to accumulate. A new astronomical object emitting regular pulses of radiation was discovered in 1967. At first there was a flurry of excitement as it was argued that the extreme regularity of the pulses indicated that they must be artefacts produced by an extra-terrestrial intelligence. They were designated by the acronym LGM, for "little green men". Sanity soon prevailed, and the radiation was explained as a beam projecting into space, focussed by intense magnetic fields surrounding a mystery object. As the object rotated, the beam swept across space, intersecting the earth regularly in an analogous fashion to a lighthouse beam sweeping across a ship.

Similar objects were discovered regularly in the following years. They were named *pulsars*. Their high speed of rotation indicated that they must be relatively small, about 20 km in diameter. It was suggested that they might be an object that had long been predicted but not yet observed: a *neutron star*.[10] The mass of a neutron star is found to be about 1.4 solar masses, which leads to a density so high that on earth one teaspoonful of neutron star material would weigh about a billion tonnes.

According to General Relativity, such extremely compact objects revolving in binary systems might be expected to perturb space–time significantly as they rotate about each other. The result should be the emission of a gravitational wave, which would carry energy away from the system, and radiate it throughout space. An example of such a binary pulsar system is discussed in Appendix 7.4. The confidence of physicists in the reality of gravitational waves therefore grew, and motivated the construction of even more sensitive apparatus in an attempt to detect these elusive ripples in space–time on their passage through the earth.

Besides binary pulsar systems, other stronger sources of gravitational waves were known to occur, and offered more hope of detection. For instance, nearby exploding supernovas might be detectable. However, such events are very rare; the last two supernova explosions in our galaxy occurred four centuries ago. Another possibility consisted of two black holes orbiting about each other, and passing into the stage of a final merger. However, the question was still open on whether such systems even existed.

---

[9]"Sceptic" may be a pejorative term in some circles, but scepticism is important in science and is an example of Occam's Razor in practice.

[10]Neutron stars were predicted by Walter Baade and Fritz Zwicky at the meeting of the American Physical Society in 1933, less than 2 years after the discovery of the neutron.

Fig. 7.11 Aerial photograph of the Virgo Detector near Pisa, Italy. Image courtesy of Virgo Collaboration (Image: The Virgo collaboration/CCO 1.0 https://www.ligo.cal tech.edu/image/ligo20170927e)

To pursue this question, extremely sensitive detectors were developed, taking the technology off in a new direction. A detector was constructed by splitting a beam of light from a laser into two components, then sending these beams along paths several kilometres long, perpendicular to each other. An incoming gravitational wave would be expected to change the length of one path slightly, compared to the other.

Two such LIGO (Laser Interferometer Gravitational-Wave Observatory) detectors were built at Hanford, Washington and Livingston, Louisiana in the U.S.A. They are separated by 3000 km, and are used in coincidence with each other. To be accepted as a real event, a gravitational wave has to be recorded at both detectors almost simultaneously.[11] By measuring the slightly different times of arrival of the signal at the two detectors, the direction in space of the source can be estimated. A third detector, called Virgo, was constructed near Pisa in Italy (see Fig. 7.11). A fourth detector has been completed in Japan

---

[11] For a GW travelling at a velocity equal to that of light, its times of arrival at the two sites could differ by up to 0.01 secs, depending on its angle of arrival with respect to the line joining Hanford and Livingston.

and should be operational in 2020, and a fifth is under construction in India, and expected to join the network in 2025. These detectors will improve the accuracy of source location.

On September 14, 2015, half a century of perseverance since the time of Weber finally paid off, when a gravitational wave was detected at the Hanford and Livingstone observatories [8]. The Virgo observatory was offline at this time undergoing upgrades. The signal arrived at Livingstone 0.007 s before Hanford, indicating the source lay in the southern hemisphere.

Great care was taken to eliminate any possibility of error. The wave was interpreted as arising from the final fraction of a second of the merger of two black holes into a single, more massive, spinning hole, and provided the first evidence that such phenomena exist. LIGO scientists estimated that the black holes had masses of approximately 29 and 36 times that of the sun, and the event occurred 1.3 billion years ago.

Since that historic date, detections have become almost commonplace. The next twist in this exciting tale occurred on August 17, 2017 [9]. This time it was not merging black holes that were detected, but two neutron stars spiralling in and colliding with each other, only some 130 million light years away. The masses of the neutron stars were 1.1 to 1.6 solar masses. (A similar event was detected on 25th April, 2019.)

What was particularly exciting about this event was that optical, X-ray, Gamma ray and radio telescopes were quickly pointed in the direction of the source, as indicated by the LIGO/VIRGO measurements, and within 12 h had located it, a fireball at the edge of a galaxy 130 million light years away. Figure 7.12 shows a photograph of the galaxy taken by the Hubble Telescope. The source is clearly visible. Over the 6 days following the photograph, the source faded quickly away.

Bringing so many different types of observation platforms to bear on the same source enables details of competing gravitational theories to be checked against each other. The fact that electromagnetic signals (gamma-rays) and gravitational waves arrive at the same time, (i.e. GW travel with the velocity of light) is in agreement with Einstein's theory and excludes some of the alternative contenders for a theory of gravity.

With the development of further, more powerful, LIGO observatories, gravitational waves have now emerged as a possible means of exploring the physics of the early years of the universe, at a time period less than 380,000 years after the Big Bang, when electromagnetic radiation cannot penetrate (see Chap. 11). This new field of astronomy has been named *multi-messenger astronomy*, as it utilises widely different technologies. The future will be watched closely to see what it brings.

Fig. 7.12 Hubble telescope photograph of Galaxy NGC 4993 on 22nd August, 2017. The box in the image (and enlarged inset) shows the optical flare from two colliding neutron stars. Within 6 days, the flare had faded to invisibility. Image: NASA/Swift (The image was released by NASA/Swift. https://www.nasa.gov/press-release/nasa-mis sions-catch-first-light-from-a-gravitational-wave-event (accessed 2020/05.27))

To end on a human note, this recent discovery of gravitational waves is a triumph of perseverance for many, now elderly, physicists, who have spent their whole professional lives on what many believed would be a fruitless search.

References

1. Newton I (1979) Opticks. Dover, New York
2. Soares DSL (2009) Newtonian gravitational deflection of light revisited. https://arxiv.org/pdf/physics/0508030.pdf. Accessed 26 May 2020
3. Ellis R et al (2009) 90 years on—the 1919 eclipse expedition at Príncipe. Astron Geophys 50(4):4.12–4.15. https://doi.org/10.1111/j.1468-4004.2009.50412.x. Accessed 26 May 2020

4. https://physics.aps.org/synopsis-for/https://doi.org/10.1103/PhysRevLett.110.101101. Accessed 27 May 2020
5. https://www.livescience.com/27811-creating-mini-black-holes.html. Accessed 27 May 2020
6. Einstein A, What is the theory of relativity? (German History in Documents and Images, vol 6. Weimar Germany, 1918/19–1933 (November 28, 1919))
7. Popper DM (1954) Red Shift in the Spectrum of 40 Eridani B. Astrophys J 120:316
8. https://www.ligo.caltech.edu/news/ligo20160211. Accessed 27 May 2020
9. https://www.ligo.caltech.edu/news/ligo20171016. Accessed 27 May 2020

8

The Most Accurate Theory in Physics

"It has long been an axiom of mine that the little things are infinitely the most important." – Sherlock Holmes [1].

8.1 The Storm

The Blue Morpho, or *Morpho peleides*, is an exquisite tropical butterfly inhabiting the rainforests of South America. Its wings gleam with a sapphire iridescence, a result of the diffraction of the forest light by the millions of tiny scales covering their surfaces. Nature has deemed that a creature so beautiful should not be doubly gifted, and to compensate for its splendour has cut its lifetime short, from egg to adult spanning a mere 115 days.

Beauty, however, has little impact on predators, such as the jacamar or the flycatcher, two of the insectivorous birds who make the forest their home. To them, the blue colour is a beacon advertising food, not unlike the neon-lit logo of a fast food chain.

And so it was that on a still afternoon in the year of 1880, in what is now known as the Common Era, a lone Blue Morpho, in the prime of its short existence, was forced into a fight to the death with a hungry marauder, many times its body weight. A

R. Barrett and P. P. Delsanto, *Don't Be Afraid of Physics*, https://doi.org/10.1007/978-3-030-63409-4_8

brave little butterfly, it fought valiantly, flapping its blue wings in a frenzy of frustration and indignation. "Why now? What could possibly be the purpose of such a short life?" were its dying thoughts.

However, if it could have foreseen the consequences of that afternoon, the Morpho might have railed less at the injustice of fate. Some natural systems are so delicately poised that a minute perturbation to one small component of them can start a cycle of ever-increasing changes, until the consequences become devastating [2]. In this instance, the death throes of the beautiful insect triggered tiny air currents that, ever so marginally, perturbed the finely-balanced prevailing meteorological conditions. Winds changed their directions slightly, isobaric weather patterns around the earth were influenced, and inexorably the conditions began to form for the development of a prodigious tempest in an altogether different part of the world.

The effects extended as far as Foxhill, a hamlet south of Nelson in the north of New Zealand's South Island, where a clap of thunder shattered the late evening tranquillity. Martha Rutherford raced to the linen cupboard to haul out an armful of bedsheets. She signalled to her daughters to help her, and before long all mirrors in the house had been safely shrouded against lightning, as was the custom of the time in a thunderstorm. The boys closed the curtains on the windows, and securely stowed silverware and other metallic objects in drawers and cupboards. There was little prospect of further sleep, as the violent electrical discharges gathered in fury outside.

Martha hated storms, but her husband, James, pooh-poohed her, and led his large family, or those who were not too afraid, out onto the verandah to watch the fiery display, and listen to the sounds of the tempest. James was a Scottish wheelwright who had immigrated with his family to New Zealand in 1842. He was a man with an interest in nature, and he demonstrated to Ernest, his nine-year-old son, how to estimate the distance to the lightning strikes by counting the seconds between each flash and the following thunderclap. He explained the difference between sheet lightning, fork lighting and ball lightning.

Before long, the flashes and thunderclaps became simultaneous. A fork of lightning struck a tree in a neighbouring field and drove the livestock into a panic. They careered madly around their yards, colliding with fences and gates. The sheer ferocity of nature's display made a deep impression on the young Ernest.

Much damage was done that night. Stock lay dead from lightning strikes and many trees were uprooted. However, as is the lot of country folk the world over, disasters were put aside, the routine of daily life resumed, and memories faded.

For one small boy, however, the storm was a watershed, and the path of his life took a new direction. Ernest became introverted and thoughtful. Sometimes he would sit at the table with knife and fork in hand, and stare into space, contemplating some problem of a scientific nature. His siblings drew their father's attention to him with their friendly banter: "Dad, look at Ern. He's away again."

In the following years, the young Rutherford developed into his country's greatest scientist, and departed from the land of his birth, as so many of his generation did, to pursue his career overseas. There he achieved the Nobel Prize, the greatest of all accolades, and became recognised as "the father of nuclear physics." His work motivated others, and in the dark period of the Second World War, it was turned to the construction of the ultimate weapon.

And so it came to be that on the 6th August, 1945, sixty-five years after the death of that lonely butterfly in an Amazonian rainforest, an American B-29 bomber took off from Tinian, one of the Northern Mariana Islands, and set a course for Hiroshima. This time the fiery tempest that was unleashed changed not just one man's destiny, but the course of human history for all time.

* * *

In this Chapter, we see how Rutherford's experimental work set the basis for a deep understanding of the interactions inside the atom, and led eventually to one of the most accurate theories in the whole of physics, a theory with a precision equivalent to predicting the distance from the earth to the moon with an accuracy of half a millimetre.[1] And yet, hidden away in the

---

[1] We should mention here that the accuracy with which General Relativity has been verified is continuously improving, especially with regard to the observation of binary systems (see Appendix 7.4) and may well rival that of Quantum Electrodynamics (QED) to be discussed later in this Chapter.

very core of this masterwork is a *caveat*, an essential, rather dodgy, sleight-of-hand, to remind us that we are but human, and our knowledge is therefore imperfect.

First, however, it is necessary to refresh and extend our understanding of two concepts that we have encountered already in previous chapters: particles and fields.

8.2   Particles

In Chap. 5, we have seen how the idea that matter in the universe is composed of elementary particles, called atoms, has been around at least since the time of the early Greeks. By the start of the 20th Century, it was recognised that electromagnetic radiation also exhibits corpuscular properties in some circumstances. Any doubt about the atomic structure of matter—and there had been many doubters because of the difficulty of observing atoms—had been dispelled by Einstein's paper in 1905 explaining Brownian motion (see Chap. 5).

The 19th Century had also seen the discovery of the electron, the fundamental particle of electricity. In 1897, the British physicist J. J. Thompson reported in *The Philosophical Magazine* experiments on the strange *cathode rays* emitted from a heated metal in a vacuum. His ingeniously designed experiments used deflections of the cathode rays by electric and magnetic fields to verify that the rays were indeed comprised of *electrified particles*, now known as electrons, and went on to determine the ratio of the electric charge of the particles to their mass.

A number of other rays were also discovered towards the end of the 19th Century. These were X-rays, produced by Wilhelm Röntgen in 1895, and the alpha, beta and gamma rays emitted during the radioactive decay of uranium, and discovered in 1899 and 1900 by Henri Becquerel, Paul Villard and Ernest Rutherford, working independently. Further experiments to determine the electric charge-to-mass ratio of the alpha and beta particles found that the positively charged alpha particle was most probably a helium atom with its two electrons stripped away, and the negatively charged beta particles were electrons. Gamma rays were a very energetic form of electromagnetic radiation. Both X-rays and gamma rays, and also visible light, heat rays (infra-red radiation) and *chemical rays* (ultra-violet radiation), are now known to be made of *photons* of different frequency (see Fig. 5.1).

So at the end of the 19th Century there was a widespread belief that apart from a few wayward examples (electrons, and the particles emitted in radioactive decay), the fundamental building blocks of matter are atoms. Such a belief did not survive for very long. Indulging in a pursuit that is fairly common among curious children, physicists began smashing objects together to see what lay inside.

Ernest Rutherford, together with Hans Geiger and Ernest Marsden, bombarded gold atoms in a thin foil with alpha particles, and observed that a significant number of these positively charged alpha particles were deflected at large angles, some even travelling back towards their source. Rutherford attributed the deflection to electrostatic repulsion. However, the gold atom is electrically neutral, so large deflections were not to be expected. A detailed mathematical analysis showed that the observed deflections (called scattering) could be explained if the gold atom had a small, positively charged *nucleus* at its core, surrounded by a diffuse cloud of negative charges (presumably electrons), to maintain its electrical neutrality. The scattering is illustrated in Fig. 8.1. This picture of the atom became known as the *Rutherford Model*.

This model immediately leads to a novel question: what is the atomic nucleus made of? From a series of experiments, Rutherford inferred that the nucleus was constructed of two new particles: *protons* and *neutrons*. The proton is the positively-charged nucleus of the hydrogen atom; the neutron is a very similar particle to the proton, but is uncharged. Its existence was confirmed in 1932 by James Chadwick. In the same year, the positive electron, or *positron*, was discovered by Carl Anderson. This was another particle whose existence had been postulated, in this case by theorist Paul Dirac, before its actual discovery.

Dirac developed a theory that integrated the quantum mechanics of Schrödinger and Heisenberg with Einstein's special relativity for a charged particle. He was struck by the *beauty*, which for a physicist usually means

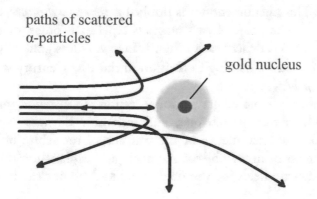

Fig. 8.1 Trajectories of alpha particles scattered from a gold nucleus in the Rutherford Model of the atom

symmetry (and relative simplicity), of his equation. He noticed that the equation would work just as well for positively charged particles as for the negatively charged electrons. He then proposed that for every particle there exists a corresponding *anti-particle*, with nearly identical properties to the original particle, except that it has the opposite electric charge. This led to the concept of *antimatter*, which consists of anti-particles in the same way that matter is constructed from ordinary particles.

As a result, in the first few decades of the 20th Century physicists (and chemists alike) had arrived at a fairly consistent picture of the structure of matter. All substances were composed of chemical compounds, whose basic building blocks were molecules, which were themselves constructed from an arrangement of atoms. The simplest possibility was a material (known as an *element*) constructed from only one type of atom. The atoms were not indivisible, as had been assumed by Democritus and Dalton, but were composed of a nucleus surrounded by a cloud of electrons. The nucleus was made up of neutrons and protons. It appeared that protons, neutrons and electrons, together with photons, were all that was necessary to explain the world around us. Physicists at this time had reason to feel well-satisfied, even smug with their achievements.

However, this happy state of self-congratulation lasted only for a short time before other particles began to be discovered, demanding an explanation for their existence. Also, one would expect the positively charged protons to repel each other electrostatically and fly apart, rather than suffer confinement within the bounds of the nucleus, unless there were some stronger, as yet unknown, nuclear forces binding them together.

Before addressing these problems, however, we must digress to explore how particles interact with each other. Besides the direct collision of one particle with another, it was clear that particles must also interact in a more indirect manner. This phenomenon was dubbed *action at a distance*, and it takes place when the like poles of two magnets repel each other, or two electrically charged objects deflect each other. It involves ideas gradually developed over many years, until coming to fruition in the 19th Century with the new concept of a *field*.[2]

We have already explored gravitational "action at a distance", aka the gravitational field, in the last Chapter, and seen how Einstein explained this phenomenon as a consequence of the warping of the fabric of space–time by the large mass of heavy objects. An analogous explanation does not exist for electromagnetic fields, or the fields produced by nuclear forces. This is

---

[2]Michael Faraday coined the term "field" in 1849. A field can be defined as the region in which the effect of a given force is felt. It extends throughout a large region of space.

one reason why it has proved so difficult to unify Quantum Mechanics and the Theory of Relativity into a *theory of everything* (TOE), describing all the fundamental natural forces.

8.3   Fields and Relativistic QM

Action at a distance was never a concept embraced readily by physicists. Newton's theory of gravity was criticised by Descartes and Leibnitz on the grounds that he had not explained the nature of gravity, which appeared in his writings as almost supernatural. A quarter of a century after the initial publication of his ideas in *Principia Mathematica*, Newton addressed these criticisms. He admitted he did not understand how gravity exerted its pull on objects across empty space.

Similar conceptual difficulties arise when considering the mutual interaction of magnets and of static electric charges. Gradually, in the nineteenth century, these *non-local* influences were attributed to fields which surround the particles. For instance, an electric field emanating from a charged particle is capable of influencing another charged particle some distance removed. Conversely, the field produced by the second particle can influence the first, resulting in a chicken-and-egg dilemma.

Now let us suppose that the first electrically charged particle is moved. How does the electric field in the vicinity of the second particle behave? Does it change instantaneously as the first particle moves, or is there a time lag? These questions were addressed by Gauss and Faraday, before being resolved by James Clerk Maxwell, as we have seen in Chap. 5, in a seminal paper in the second half of the 19th Century.

Maxwell found that not only did the electric field propagate out from a moving electric charge with a definite speed, but the moving electric charge generated a *magnetic* field, which also propagated. These propagating intertwined fields became known as *electromagnetic radiation* and the study of their properties was called *Electromagnetism*.

Maxwell went even further: he was able to predict the speed of propagation of the electromagnetic radiation from two independent physical constants which could be measured in the laboratory. The value he obtained for this speed was close to another well-known physical constant, the speed of light. Not believing this to be just a coincidence, he then made the bold assertion that light was actually a form of electromagnetic radiation. His work united the three separate disciplines of electricity, magnetism and optics under the banner of electromagnetism, and became one of the bedrocks of what is now known as *classical physics*.

We now turn our attention to a phenomenon known as *the line spectra of atoms*, whose explanation provided some of the most convincing evidence for the veracity of Quantum Mechanics. In the non-relativistic Quantum Mechanics of Schrödinger, the energies of the electrons in the cloud surrounding the nucleus of an atom may only have certain particular values, or *energy levels*. The value of the energy in these levels is obtained from the solution of the fundamental equation of QM, the *Schrödinger Equation*. Electromagnetic radiation is emitted in the form of discrete frequencies, when electrons transition from a higher to a lower energy state. This is quite different from the predictions of classical physics, where a spectrum continuous across all frequencies, is expected.

In Fig. 8.2, we see an example of this effect, when common salt (sodium chloride) is sprinkled into a gas flame. The flame changes to a bright yellow as electrons in the cloud surrounding the sodium nucleus are excited to a higher level, and then fall back, with the emission of light of two characteristic frequencies close together in the yellow region of the visible spectrum.

This situation is shown schematically in Fig. 8.3. As an electron falls from a higher to a lower energy level, it emits a photon with an energy equal to the difference between the two levels. This is the original "quantum leap".

Now, just as water runs down a hill to take up a position of low potential energy, so electrons in the electron cloud would be expected to cascade down into the lowest possible energy level. The reason this does not happen is embodied in an *Exclusion Principle* proposed by Wolfgang Pauli in 1925. This

Fig. 8.2 Common salt (sodium chloride) sprinkled into a gas flame produces a bright yellow flare, characteristic of the decay of excited sodium atoms

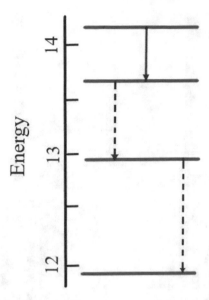

Fig. 8.3 A schematic picture of electron transitions between energy levels in an atom. Note that not all transitions (e.g. the dashed ones) are possible, as some are forbidden by the Pauli Exclusion Principle (see text)

principle states that two identical electrons cannot occupy the same energy level simultaneously.

There are rules arising from the Pauli Exclusion Principle for the way that electrons may place themselves systematically in the atomic energy levels. We need not go into these here. However, suffice it to say that these rules explain the Periodic Table of the Elements, discovered empirically by Mendeleev in the 19th Century, which predicts the properties of the chemical elements, including those of some elements that had not been discovered at that time.

When a sodium atom is placed in a magnetic field, the spectral lines split, showing fine structure that is unpredictable from Schrödinger's equation. Figure 8.4 is an original photograph of line spectra taken by Pieter Zeeman in 1897, showing how the spectral lines in the upper half of the photograph split into the multiple lines in the lower half, when a magnetic field is applied.

When Dirac's extension of Schrödinger's equation to include relativistic effects was applied to this problem, it was startlingly successful. His equation explained the *fine structure* in the line spectra even when the atoms were placed in a magnetic field. As a bonus, it predicted the intrinsic angular momentum carried by every electron. This quantity was dubbed the electron's spin, in analogy with the angular momentum of a spinning ball.

There is a danger, however, in using models based on our everyday experience in domains that are far removed (in terms of size and velocity) from our

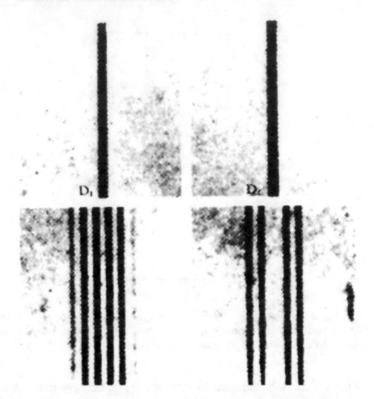

Fig. 8.4 An example of the fine structure produced in two spectral lines from sodium by the application of a magnetic field. The upper half of the photograph shows two lines when no magnetic field is applied, and the lower half demonstrates the appearance of additional lines when it is applied. Image in Public Domain (Image in Public Domain (due to age): https://commons.wikimedia.org/wiki/File:ZeemanEff ect.GIF (accessed 2020/9/4))

daily life. It soon became apparent that this classical picture was an inaccurate description of the electron. Attempts to measure the size of the electron revealed that if the electron were a spinning ball, its surface would have to be moving faster than the velocity of light to produce the observed intrinsic angular momentum. This would be in conflict with the Theory of Relativity. The electron is now believed to contain no internal structure, and have *zero size*: it is, in fact, the archetypal mathematical point, albeit carrying angular momentum, electric charge and mass. In this sense it is truly a *fundamental* particle.

How a quantity with zero volume can be massive is a question that our everyday experience does not equip us to answer. In the quantum world, we need to relax and draw inspiration from Lewis Carroll's *Alice in Wonderland*, where the Cheshire cat vanishes, leaving behind a disembodied grin (see

Fig. 8.5). *"Well! I've often seen a cat without a grin,"* thought Alice; *"but a grin without a cat! It's the most curious thing I saw in all my life!"*.

In the preceding paragraphs, we have given names to three intangible quantities (i.e. angular momentum, electric charge and mass), as if by so doing we could gain an understanding of what they are. In Chap. 2 we have discussed the difficulties associated with a search for understanding, and what we, as physicists, mean by this term. Angular momentum, electric charge and mass are found in physical equations that accurately predict the results of experiments and observations spanning the full gamut of spatial dimensions, from the smallest fundamental particles to clusters of galaxies spread throughout the universe. This wide-ranging predictive capability is an example of what a physicist means by understanding.

So if the Schrödinger Equation has been so successful at explaining the electronic structure of atoms, how does the Dirac Equation fare, including as it does aspects of the Theory of Relativity? We have already noted its success in explaining the mysterious fine structure that occurs in energy levels in the presence of a magnetic field.

Another basic difference between the predictions of Dirac's theory and Schrödinger's non-relativistic QM became immediately apparent: the Dirac

Fig. 8.5 The Cheshire cat fades away, leaving behind only his grin. *Alice's Adventures in Wonderland* by Lewis Carroll, 1865. Image by Sir John Tenniel (Image: public domain due to age https://commons.wikimedia.org/wiki/File:Alice_par_John_Tenniel_24.png (accessed 2020/05/30))

theory allows negative energy solutions. But what does this mean: an electron with negative energy? It must have been tempting at first to simply dismiss these solutions as non-physical, and sweep them under the carpet. Dirac, however, was of a different mind. He decided to assume that negative-energy solutions were real and explore the implications of this assumption.

Firstly, he investigated what would be the effect of the energy carried by a photon impinging on the electronic system described by his equation. Non-relativistic QM allows an electron to be knocked from a lower to a higher level, or even out of the atom altogether. The latter process, where atomic electrons are completely stripped from atoms, is known as ionization. These two phenomena are also possible in Dirac's theory; it would not be a tenable physical theory if this were not the case.

However, in addition, Dirac interpreted the vacuum as comprising a veritable sea of negative electrons, known as the *Dirac Sea*. The existence of such a sea of negative energy electrons allows for another possibility: that an electron can be excited out of this sea into the positive energy region, leaving behind a *hole*. This hole, called a *positron*, bears all the characteristics of an electron, except that it is positively charged. To an experimenter bombarding matter with photons in the laboratory, this phenomenon will manifest itself by the sudden appearance of two particles: a conventional electron and a positron. The process is called *pair production*, and is illustrated schematically in Fig. 8.6.

We have now come to the end of the road for the Dirac equation. It was designed for, and was brilliantly successful at describing the behaviour of a single electron. However, to fully account for pair production we need a

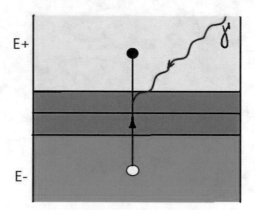

Fig. 8.6 Pair Production: An incoming photon (γ) excites an electron out of the negative energy (blue) "Dirac Sea", leaving behind a hole (positron) and a positive-energy free electron

theory that can describe the electron–electron interaction, and the electron-photon interaction. Such a theory, called Quantum Electrodynamics (or QED for short), will be outlined in the next Section. It is very relevant, even in view of our simplified approach in this book, for three reasons: first, because it is arguably the most precise theory ever to be developed in any field of science, second because it can be applied to all interactions in nature except gravity and radioactivity, and third because it has been used as a template for another theory, the Standard Model of Fundamental Particles, which is currently the most widely-accepted explanation for the "*zoo*" of newly discovered particles, that we shall encounter in Chap. 9.

8.4   Quantum Electrodynamics (QED)

QED owes its development largely to three physicists, who shared the 1965 Nobel Prize for their "*fundamental work in quantum electrodynamics, with deep-ploughing consequences for the physics of elementary particles*". These scientists, who worked independently, were Sin-Itiro Tomonaga, Julian Schwinger and Richard P. Feynman, and their approaches to the problem were quite different.

It is not uncommon in physics that different ways of tackling a problem lead to the same answer. For instance, Newton and Lagrange approached classical mechanics very differently. Newton employed forces, velocities and acceleration, which are *vectors*,[3] whereas Lagrange used a sophisticated mathematical technique, known as the Calculus of Variations. Almost two centuries later, Quantum Mechanics was formulated using differential equations by Schrödinger and matrices by Werner Heisenberg. Both of these apparently disparate methods have been shown to be mathematically equivalent. In the case of QED, the approach developed by Feynman is the most transparent and will be discussed in the following.

Let us begin with a simple illustrative example: the reflection of light from a mirror (see Fig. 8.7). In the classical representation of Maxwell, light is a wave and its reflection is no more difficult to explain than the reflection of ripples from the side of a pond. But, as we have seen in Chap. 5, in QM light is composed of a stream of particles (photons), whose distribution is controlled by a probability wave function.

In the upper figure we represent rays of light, transiting from the source at A and arriving at B, after having been reflected by a mirror. The ray reflected

---

[3]Vectors are mathematical entities, representing physical quantities, which possess both a numerical value (called the *modulus*) and a direction.

at C is the only one, for which the angle of incidence is equal to the angle of reflection, and is the only one we would consider in a classical physics treatment. However, the probabilistic nature of quantum mechanics implies that we cannot exclude the other rays shown in Fig. 8.7.

The lower half of the figure is a graph of the path length travelled by all rays travelling from A to B via reflection from the mirror, and holds the key to the reconciliation of the quantum and classical results. From this plot, it is clear that the ray reflected at point C on the mirror's surface has the shortest path length. Let us call this ray the *primary ray*. Rays close to the primary ray have path lengths almost equal to that of the primary ray. As we move away from the primary ray, the path length increases rapidly.

As we have seen in Chap. 5, the distribution of photons in the quantum mechanical picture is derived from the solution of a wave equation. All possible reflected paths from A to B must be considered. The paths of photons close to that of the primary ray have similar path lengths, and so their probabilities add up; in other words, these rays reinforce each other. The path lengths of photons far from the primary ray increase rapidly. The contribution from one ray in this region is cancelled by that from a neighbour with a path length one half-wavelength different from its own. The net effect is that the contributions from rays in this region to the overall probability cancel each other. The rays close to the primary ray therefore form overwhelmingly the major contribution to the total probability distribution, and rays impinging on the mirror away from C can normally be disregarded.

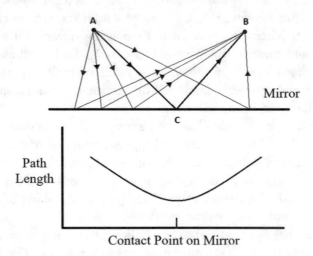

Fig. 8.7 Upper figure: rays of light travelling from A to B, after reflection from a mirror. Lower figure: the path length of the various rays, depending on their contact point with the mirror

Is there any way we can allow rays far away from the primary ray to make their presence felt? Yes, there is. We can scratch lines across the mirror with a spacing such that only those rays with a path length equal to a multiple of the wavelength of the incident light are reflected. (We are assuming here that the incident light is monochromatic.) In this case, the probabilities from these rays will add up, and all rays will make a contribution to the reflected light arriving at point B. Such a scratched mirror is called a *diffraction grating*.

We have discussed this example in some detail to make a point that is necessary in what follows: all possible paths contribute to the probability of finding a photon arriving at point B. However, contributions from particular paths often cancel each other, leaving the majority of the contribution to relatively few of the possible paths.

We now return to what is the core of QED, the interaction of electrons with each other, and with photons. In so doing we will encounter some strange effects that will stretch our credulity. Some readers may be unwilling to accept these effects because they are too "crazy". However, Richard Feynman summed up his reaction to this objection: *"It's the way nature works. If you don't like it, go somewhere else. Go to another universe where the rules are simpler, and philosophically more pleasing"* [3]. The fact is that the accuracy of QED in describing natural phenomena, with the exceptions of gravity and nuclear phenomena, is mind-boggling, and we cannot just discard it because we find its principles contrary to our preconceptions.

In the classical model, a stationary electron is pictured as an electric charge sitting in space, surrounded by an electric field. In QED the energy in the electric field results in the spontaneous creation of photons and/or electron–positron pairs. This situation is represented in Fig. 8.8.

Figure 8.8a displays the classical situation, with the electron remaining at the same position in space as time progresses. (In these diagrams, time is plotted vertically, and spatial distance horizontally, even if the axes are not specifically plotted). In Fig. 8.8b, a photon has been spontaneously produced by the electron and has been reabsorbed a short time later. (The photon's path is represented by the wiggly line). Figure 8.8c shows a more complex case where two photons have been produced, and reabsorbed. There are an infinite number of such possibilities. These diagrams are called Feynman diagrams, after their inventor.

As we saw earlier, if its energy is sufficiently high, a photon can knock an electron out of the Dirac Sea, and leave behind a hole, which is interpreted as a positron. The corresponding Feynman diagram is shown in Fig. 8.8d. The electron and positron may recombine (or *annihilate*) a short time later to produce another photon, as shown in Fig. 8.8e.

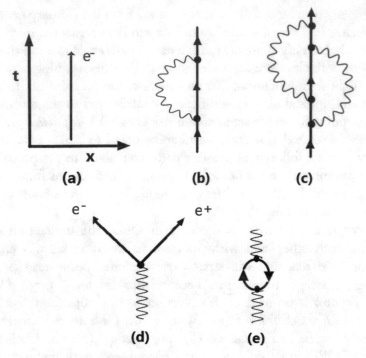

Fig. 8.8 Examples of Feynman diagrams

In Fig. 8.8e we have taken advantage of the symmetry of the laws of physics with respect to the reversal of time. If we take a video of the collision of two billiard balls on a table, it is not possible just by examining it to decide whether the video is running forwards or backwards. An analogous symmetry in the equations of QM means that we cannot differentiate between an electron moving backwards in time and a positron moving forwards. In Fig. 8.8e we have observed this convention by depicting a positron as a backwards-moving electron.

Figure 8.8 contains only a few samples of Feynman diagrams. Their number is limitless, and in analogy with the rays in Fig. 8.7, *all of them* must be considered when calculating the quantum probabilities associated with any event. This may seem an impossible task. However, as we shall see later, many of these diagrams make negligible contributions to the quantum probability, and can be safely neglected, in analogy with the neglect of outlying rays in Fig. 8.7.

To conclude, we arrive at a picture in which the space surrounding an electron is alive with photons potentially being created and absorbed all the time, and electron–positron pairs being continuously created and annihilated. The particles that are created and recombine around the original particle are known as *virtual particles*. They differ from real particles in that, as a consequence of the Heisenberg Uncertainty Principle that we encountered in Chap. 5, they do not necessarily obey conservation laws. If a particle is short-lived, the time when it is in existence can be specified quite precisely. The Uncertainty Principle tells us that in this case the particle's energy cannot be measured accurately (which is contrary to the Law of Conservation of Energy), and QM tells us that all significant graphs must be considered when taking the sum over all probabilities, not just those where energy is conserved.

Our final example of a Feynman diagram is shown in Fig. 8.9. Here two electrons pass close by each other and exchange a virtual photon. It is this process that at last explains the mystery of the electromagnetic field, allowing the interactions at a distance that have baffled physicists for centuries. Of course, Fig. 8.9 is only the simplest of a series of Feynman diagrams that must be included in the sum of probabilities to obtain a correct evaluation of the effect.

It is now time to put the theory that we have outlined to the test of experiment. Charged particles, such as the electron, which also have angular momentum, or spin, behave as though they are tiny magnets when placed in a magnetic field. Physicists say that these particles have a *magnetic moment*. This behaviour is predicted by Dirac's theory, and observed by experiment. The fact that the theoretical and experimental values for the magnetic moment agree closely was regarded as a triumph of Dirac's relativistic quantum mechanics.

However, as experiments became more accurate, a small discrepancy of about one part in a thousand emerged between Dirac's theoretical predictions and the experimental measurements. Such a difference, expressed as a fraction of the Dirac prediction, was called the *anomalous magnetic moment*, and given the symbol α_e. The current experimental value of α_e is 0.001 159 652 180 73(28). The number in parentheses is the estimated experimental error in this result. To obtain this value requires measuring the electron magnetic moment to a precision of one part in a trillion (1 in 10^{12}).

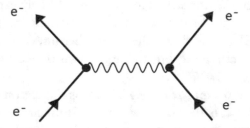

Fig. 8.9 Collision between two electrons with exchange of a virtual photon

The first improvement on Dirac's theoretical value (which was 2) for the electron magnetic moment was made by Julian Schwinger, who obtained $\alpha_e = 0.001\ 161\ 4$ in 1948. This result effectively made use of only the most important Feynman diagram. As more complex Feynman diagrams were added, this theoretical prediction became more and more refined. Remember: in principle all of the infinite number of Feynman diagrams should be included. However, the contributions from Feynman diagrams become less and less important as the diagrams become more complex, which makes it possible to terminate the summation at some stage. Currently the best analytical result for α_e is $0.001\ 159\ 652\ 181\ 643\ (764)$. The number in parentheses is the possible error attributed to neglected Feynman diagrams.

The agreement between theory and experiment here is astonishing. However, in all honesty it must be admitted that a skilled piece of legerdemain has taken place. The theoretical values depend on a fundamental physical constant known as the *Fine Structure Constant* (see Chap. 3), which, to add to the confusion, is represented by the symbol α. Physicists have not found a way to compute the value of α from first principles, and must rely on obtaining it from experimental observations, by adjusting it until the QED predictions agree with experimental measurements.

Clearly such a procedure (adjusting a constant to obtain the desired result) would not normally be acceptable, but fortunately, the anomalous magnetic moment is not the only physical quantity that depends on the fine structure constant. For instance, atom-recoil measurements give a value of α such that:

$$\overset{-1}{\alpha} = 137.03599878(91),$$

which is to be compared with the value

$$\overset{-1}{\alpha} = 137.035999070\,(98)$$

obtained from the magnetic moment experiments. (As we mentioned already in Chap. 3, the fine structure constant is usually expressed as a reciprocal.) The consistency of these two results is a demonstration of the internal consistency of QED. In addition, there are other physical quantities that can also be used to determine α, but the above two provide the most accurate values.

The low value of α (approximately 1/137) is the reason that higher order Feynman diagrams can be neglected in the summation process described above. Each Feynman diagram contains the factor α multiple times, equal to the order of the diagram. As a consequence third order diagrams have three α

factors, equal to $1/(137 \times 137 \times 137)$, or $1/2,571,353$. Orders higher than this can be safely neglected.

Before concluding this Chapter we must draw attention to a very large elephant that has been hulking around the offices and laboratories of physicists since the middle of the last century. So accustomed have they become to its presence that they pursue their studies amidst the elephant's trampling limbs and swinging trunk with scarcely a moment of concern. They have even given it a name: *renormalisation.*

The first indication of its presence came when considering the effect of virtual electron–positron pairs created in the vicinity of a real electron. The real electron should attract the virtual positrons towards it, resulting in it appearing to have less electric charge than it actually has. QED enables us to calculate how much this reduction of charge should be, and the answer turns out to be infinite. Undaunted, physicists simply propose that the unshielded electron has an infinite negative charge so that when the infinite positive shielding charge is added to it, the result is just the small negative value that the electron actually has. To sum up:

$$\infty - \infty = -1.60 \times 10^{-19} \text{Coulombs.}$$

This procedure is known as renormalisation. The electron mass also requires a similar sleight-of-hand to obtain the experimentally observed value.

We have discussed in Chap. 3 the difference between physicists and mathematicians in their attitudes to truth. No mathematician would consider subtracting two infinities from each other as anything but nonsense. To be honest, many physicists agree with them, but would be unwilling to confess it in public. Nobel Laureates can afford to be more forthright:

"I must say that I am very dissatisfied with the situation, because this so-called 'good theory' does involve neglecting infinities which appear in its equations, neglecting them in an arbitrary way. This is just not sensible mathematics. Sensible mathematics involves neglecting a quantity when it turns out to be small—not neglecting it just because it is infinitely great and you do not want it!" – Paul Dirac, Nobel Laureate, 1933 [4].

"But no matter how clever the word (renormalisation), it is still what I would call a dippy process! Having to resort to such hocus-pocus has prevented us from proving that the theory of quantum electrodynamics is mathematically self-consistent. It's surprising that the theory still hasn't been proved self-consistent one way or the other by now; I suspect that renormalization is not mathematically legitimate." – Richard Feynman, Nobel Laureate, 1965 [5].

However, once the renormalisation procedure is carried out, the agreement of experiment with the predictions of QED is so astonishingly good that there can be little doubt that QED is largely correct. The infinities seem to arise from virtual particles with very high energies. The suggestion has been made that QED breaks down at these energies, in the same way that Newtonian mechanics breaks down at speeds close to the speed of light. Time will eventually tell. In the meantime, as long as QED produces the highly accurate results that it does, it will continue to be applied unreservedly.

In the next Chapter, we will discuss the attempts to extend the methodology of QED to forces other than the electromagnetic interaction, which will lead us to the Standard Model of Fundamental Particles.

References

1. Doyle AC, The memoirs of Sherlock Holmes
2. Gleich J (1988) Chaos, Chapter 1: the butterfly effect. Sphere Books, p 9
3. Feynman R (1979) Lectures on quantum electrodynamics. Sir Douglas Robb lectures at the University of Auckland, 1979, Lecture 1. https://cosmolearning. org/courses/richard-feynman-lectures-quantum-electrodynamics/
4. Kragh H (1990) Dirac: a scientific biography. Cambridge University Press, Cambridge, p 184
5. Francis C (2018) Light after dark II: the large and the small. Troubador Publishing Ltd, p 151

9

The Standard Model of Fundamental Particles

"Who ordered that?" – a quote attributed to physicist, I. I. Rabi, on hearing of the discovery of the muon, a particle that nobody wanted.

9.1   Alice in Nanoland

Ever since her visit to Gödel-land, Alice had been restless, and unable to concentrate on her school lessons. She pined for another dose of similar excitement, so much so that her mother became concerned for her health, and decided to take her on a visit to Aunt Elspeth, the white witch of the nor'-nor'-east.

Elspeth examined her niece closely, and then reached for two bottles of pills. "Alice has my blood", she explained. "She will never be satisfied with the humdrum. She craves adventure."

"But she's only a child," objected Marigold, Alice's mother.

"These red pills will make her shrink, so that everything around appears different. I mean, really shrink! An ant will look as big as a dinosaur, and will be just as dangerous. It's totally awesome. But be careful."

"She's too little for anything like that."

"The green pills will bring her back to normal size."

"Elspeth, you're not listening to me."

© The Author(s), under exclusive license
to Springer Nature Switzerland AG 2021
R. Barrett and P. P. Delsanto, *Don't Be Afraid of Physics*,
https://doi.org/10.1007/978-3-030-63409-4_9

Marigold's protests were too late, because Alice had already snatched the red container, and began swallowing pills, one after the other, as if they were chocolates. Marigold grabbed a second red container, and then the green one, and popped the red pills in an attempt to keep up with her daughter.

At first Marigold thought that the world about her was growing. The nearby table was now so big she could no longer see over the top. Her sister was a huge giant, with a silly grin on her face, who was waving a hand as big as a side of beef at her. Only Alice seemed at all normal, except for a gleam of rapture in her eyes. She was clearly enjoying herself, more than she had done for many months.

Their shrinkage was now accelerating. Down they went through a forest that Marigold realised must be the nap of the carpet. As they got smaller, they were surrounded by minute animals propelling themselves around by various mechanisms. Some sort of microbe, perhaps, thought Marigold. When was this going to end? How many of those confounded pills had they eaten?

Finally, the shrinkage slowed and stopped. Alice took hold of her mother's hand. Her sense of adventure only went so far. "Yuk! What are those ugly things?" asked Alice, pointing at a group of large, blurry structures, just to their right.

"I don't know," replied Marigold, her voice wavering with fright.

"Molecules, of course. Don't they teach you anything in school these days?" Marigold and Alice started in unison, and turned to see a little bald-headed man, about a third as tall as Alice, staring at them.

"Who are you?" they asked, as one.

"Felix Abernethy Quantum, Ph.D., F.A.P.S., R.M.V., at your service. I can see this is your first visit to Nanoland. There's nothing to be afraid of – if you're careful. I come here all the time. It makes for a nice outing."

"Oh, really," said Alice, as she noticed a continuous flashing of ultrafast, very small particles flying in all directions. "What happens if one of those things hits us?" At that very moment, Marigold was indeed struck. "Mom, mom, are you all right?"

"I'm fine. Just a tickle. It doesn't hurt at all." She began to giggle. Moments later Alice was also hit, and then Marigold again, and soon they were both laughing themselves silly. "We should have brought the fly spray. They're peskier than mosquitoes."

"They're neutrinos," said Dr Quantum. "So small that it's hard to know if they're matter at all. And they have no electric charge. That's why they don't hurt. They can even pass right through the earth without affecting anything."

"Oh, my gosh," cried Alice, squeezing her mother's hand tightly. With her free hand she pointed to a neutrino that appeared to be undergoing some sort of change, turning into a similar, but uglier, version of itself.

"Yes, there are three different types of neutrino, and they can change from one to the other," explained Dr Quantum.

"Why only three?" asked Alice. "Why not four, or five, or a thousand?"

"Don't ask so many ..." began Marigold, before Dr Quantum silenced her with a gesture.

"Three seems to be a very special number. There are also three members of the family of leptons, three generations of quarks, and three quarks make up a baryon."

"What's he talking about, Mom?" Before Marigold could think of a suitably evasive reply, Alice was tugging her arm and pointing at a large spherical blob, surrounded by a shimmering aura.

"I see you've found a hydrogen atom," said Dr Quantum. "Those are electrons buzzing about. Careful, don't get too close now."

"That thing in the middle. I can see something moving inside it. It's pregnant."

"It's a proton, and those are quarks inside it."

Alice began to giggle. "Cottage cheese. So, it ate too much quark for dinner last night, did it?"

"No, those are a different type of quark, with funny names like up and down, strange, charm, top and bottom. And there are other little things in there as well, called gluons, because they stick everything together."

"Look, there's another one, only a bit different. And over there, a big group of them, both sorts. They're having a party, with that shimmery cloud stuff all around them."

"It's another type of atom. Carbon, I think. It has a centre made up of protons, and other particles called neutrons."

"What else is there to see?" Alice's attention was beginning to wander. After all, she was only a little girl. Suddenly there was a huge explosion, and two dazzling flashes of light sped off in opposite directions. "Wow!"

"That was an annihilation. It happens sometimes when there's antimatter about. You'd better watch out for bosons. Especially the Higgs boson. If he grabs you, you will get very heavy indeed."

"Come on, Alice," said Marigold, becoming apprehensive about their safety. "It's time we were getting back. She held up the container of green pills, but Alice showed no interest.

"Are there any things smaller than these, Dr Quantum?"

"I don't know," said the little bald man, his face reddening with embarrassment at being forced to make such an admission. "Not just me. Nobody knows."

Suddenly Marigold realised what her daughter was thinking. She knew that cheeky expression only too well. "Alice, no. Stop. You mustn't. Alice!" Too late, she grabbed at Alice's hand. It was empty, and the little girl was already shrinking, on her way to another domain, where not even Dr Quantum, with all his great scientific knowledge, could tell her what to expect.

* * *

If we leave Alice, and her reluctant mother, chasing excitement in their search for the ever smaller, we must realise they are only emulating the physicists of the 20th Century, whose penchant for the infinitesimal extended the range of physics deep into the subatomic domain. In so doing, they opened up a new world of great beauty, and considerable complexity, which we shall explore in the remaining Sections of this Chapter.

9.2   The Particle Zoo

As we have seen in Chap. 8, at the end of the 19th Century, physicists were quite happy with their understanding that, together with electrons and photons, the world was comprised of atoms, bound into molecules, and that these molecules made up all of the vast variety of matter that surrounds us. There were those, such as Rutherford, who were dissatisfied that so many different types of atom existed, and believed that perhaps atoms themselves might be composed of something smaller. Rutherford's experiments, described in Chap. 8, showed that atoms were indeed constructed of neutrons, protons and electrons, and it was believed that at last the fundamental particles had been obtained. Everything of importance had been learned, or so they maintained.

Perhaps this arrogance was an example of hubris, an offence in Greek mythology that is a challenge to the gods, for it certainly brought a swift retribution down on the heads of the perpetrators. Within a decade or two, this cosy little clique of four fundamental particles (including the photon) had to be expanded to include a plethora of new arrivals, some coloured and flavoured, some charmed and others downright strange. (This may seem an unusual collection of adjectives to associate with particles, but as we shall see in later Sections, physicists working in this field love to choose curious names.) So widespread had this influx become, rather like a horde of gate-crashers storming a dinner at the Athenaeum Club, that Willis Lamb joked in his 1955 Nobel Lecture: "*the finder of a new elementary particle used to be rewarded by a Nobel Prize, but such a discovery now ought to be punished by a 10,000 dollar fine.*"

The term *Particle Zoo* was coined to describe the new state of affairs. As a consequence of the random order in which the discovery of these new particles occurred, we will abandon here any attempt at chronology, and instead arrange the particles *post factum* into groups depending on their nature.

9.3 The Four Forces

We saw in Chaps. 7 and 8 that two of the fundamental forces of nature, gravity and electromagnetism, can be explained, the former by Einstein's theory of Relativity, and the latter by Quantum Electrodynamics. These forces described all of the known physics up until the time Rutherford and his co-workers upset the classical applecart with their discovery of the atomic nucleus. Their experiments showed that these nuclei are comprised of electrically charged protons and uncharged neutrons, both with masses approximately 1837 times that of the electron.

The protons, being positively charged, experience a strong mutual electrostatic repulsion. Some new force must therefore be overpowering this repulsion and holding the nucleus together. This force was given the somewhat unimaginative name, especially considering the flamboyant nomenclature prevalent today in particle physics, of *strong nuclear interaction*. The natural question that immediately follows is: "what produces this interaction?" We saw in Chap. 8 how the electromagnetic interaction is generated by the exchange of photons. In analogy, Japanese physicist, Hideki Yukawa, proposed in 1935 that the strong interaction was the result of the exchange of as yet undiscovered particles between the nucleons (i.e., between neutrons and protons).

However, a major difference exists between the electromagnetic and strong nuclear interactions. The fact that the nuclear force is not observed in everyday phenomena, despite the great strength it must have to hold the nucleons together, is evidence that it has a short range. In the quantum theory of fields, the mass of a carrier particle associated with a force is related to the range of that force. (See Appendix 9.1.) The requirement that the new force be of such short range, i.e. of the order of a few femtometres[1], implies that, unlike the photon, the carrier particle will not be massless. Indeed, its mass could be predicted reasonably accurately. Yukawa named the hypothetical particle a "meson" (which is Greek for "intermediate") because its mass was expected to lie between the mass of an electron and that of a proton.

In Chap. 8 we introduced the concept of an intrinsic angular momentum, or spin, associated with fundamental particles. As Planck found to be the case for energy, this angular momentum is quantised, which means it comes in multiples of an elementary basic unit. (It need not concern us here what the actual value of this unit is[2]; suffice it to say that it is *very* small.) Experimental physicists, excited by the new field opening before them, were soon busy populating the particle zoo. The new particles they discovered could be divided into two classes, known as *bosons* and *fermions*, differentiated by the value of their spins. Bosons, named after Indian physicist Satyendra Nath Bose, possess a spin of integer or zero units. On the other hand, Fermions, named after Italian physicist Enrico Fermi, possess half odd-integer (i.e. 1/2, 3/2, 5/2, etc.) units of spin. As we shall see later, this difference results in some important differences in the properties of these particles. Armed with this information, experimentalists began combing cosmic ray data for evidence of the proposed meson.

One can imagine the excitement when a new particle, dubbed the mu meson, was discovered by Carl Anderson and Seth Neddermeyer at Caltech in 1936. (Carl Anderson also discovered the positron, so if Willis Lamb's suggestion above had been in effect, he would have become severely impoverished.) The elation soon turned to disappointment when the newcomer turned out

---

[1] 1 femtometre (fm) = 10^{-15} metres. For comparison, the proton radius is 0.833 fm.

[2] In QM, angular momentum, including spin, is quantised, and can only have distinct values in terms of $\hbar$, where $\hbar$ is Planck's constant, h, divided by 2π.

to be a fermion, not a boson, as anticipated from Quantum Field Theory. In fact, the particle appeared not to participate in the strong interaction at all. It therefore lost its status as a meson, and led to the disgruntled comment by Rabi at the head of this Chapter. It is now called the *muon*. The pi meson, or *pion*, which is now recognised as the real Yukawa particle, was discovered in 1947 by Cecil Powell. As a consequence, Yukawa was awarded the Nobel Prize in physics in 1949, and Powell joined him a year later.

We shall leave Yukawa and Powell celebrating their triumphs, and digress for a few paragraphs to discuss briefly the fourth of the forces occurring in nature, and its consequences for the field of particle physics. The *weak interaction* was originally proposed in 1933 by Enrico Fermi to explain beta decay, which is the emission of high-speed electrons from an atomic nucleus. It is also the force that is responsible for the decay of other fundamental particles, e.g. the decay of a free neutron into a proton, an electron and a neutrino:

$$n \rightarrow p + e + \nu.$$

(We shall see later that the particle emitted in this decay, for reasons of symmetry, is actually an anti-neutrino, the anti-particle of the neutrino.)

The *neutrino*[3] was originally proposed by Wolfgang Pauli when he realised that without some new particle to carry away excess angular momentum in beta decay, conservation of angular momentum was not possible. The neutron is a fermion, with a spin of ½. The proton and electron are also fermions with spins of ½. Without the presence of a neutrino (a fermion with a spin of ½) it would not be possible to balance the angular momenta before and after the decay process. Pauli chose what he considered the lesser of two evils. Rather than violate the law of conservation of angular momentum, he invented a new particle, a fermion with zero, or near zero, mass. Not that his action left him with a clear conscience: "*I have done a terrible thing,*" he admitted. "*I have postulated a particle that cannot be detected.*" He was, however, wrong: the neutrino *has* since been detected, and about 100 trillion of them pass through our bodies every second.

As is often the case in physics, the discovery of a new entity can have implications that no one has foreseen. The weak interaction was found to have a unique characteristic that differentiates it from the other three forces in a way that was not detected, and largely not even suspected, for two decades.

In the last Chapter, we stated that the laws of physics have a symmetry with regard to the reversal of time. If we watch a video of two balls colliding

---

[3]The name "neutrino" is Italian for "little neutral one". Perhaps Pauli thought the Italian had a better ring to it than "kleines Neutralteilchen", the German equivalent.

on a billiard table, we cannot tell if the video is running forwards or backwards. We also cannot tell whether the video has been left-right reflected, i.e. whether we are actually watching a mirror image of the collision. This is because all the laws of classical physics are symmetric with respect to these types of interchange.

This symmetry can manifest itself in many ways. For instance, suppose we wish to communicate to another intelligence in a faraway galaxy, and tell them that we drive on the left-hand side of the road[4], how can we do it by words alone? Concepts such as left or right are mere conventions, and classical physics does not offer us any experiment that we could ask our far-off friends to perform which might help them distinguish left from right. The weak interaction, however, provides us with a solution. An experiment to measure the spin of the electrons emitted in beta decay, a process that involves the weak interaction, will show an asymmetry that enables us to define the left and right directions unambiguously [1].

It is also interesting to note that complex biological molecules have a left-handedness, which may be a consequence of their all being descended from an original left-handed molecule aeons ago. Chemicals manufactured in the laboratory normally have equal numbers of left and right-handed molecules. It is worth debating whether extra-terrestrial life, if we discover it sometime in the future, will display a similar left-handedness in its constituent molecules.

As its name suggests, the strength of the weak interaction is much less than that of the strong interaction, by a factor of approximately 10^{-6}. (Because the weak force decays much more quickly with distance than the strong force, this factor is hard to define precisely.) It is also only about 10^{-4} times as strong as the electromagnetic interaction. Its range is of the order of 10^{-18} m, which is 1/1000 that of the strong force.

Following the same methodology used by Feynman in QED, and by Yukawa for the strong force, Sheldon Lee Glashow, Abdus Salam, and Steven Weinberg in the 1960s independently proposed a scheme for the weak force, analogous to QED. However, a particular symmetry requirement could only be met if the electromagnetic force were also included into their scheme. The result is a composite *electroweak interaction*. This development generated great excitement at the time, as it was seen as the first step towards a Theory of Everything, an overarching theory, uniting all four forces into one. This was the goal pursued in vain by Einstein for the latter half of his life. It seemed the journey he had been seeking had at last begun, and excitement was in the air.

---

[4]Well, some of us do, anyway.

Four massless carrier particles are required for the new interaction. Three of them were given the name *Intermediate Vector Bosons* and are represented by the symbols W^+, W^-, and Z^0. The fourth is the photon. A problem immediately arose in that the short range of the weak interaction implied that the carriers W^+, W^-, and Z^0 had to be very massive, of the order of 650 times the mass of the pion. Some mechanism is therefore required that gives mass to the three intermediate vector bosons, but not to the photon. This process was resolved by the assumption of another unseen field, the Higgs field, that is assumed to pervade all space. Interaction of the Higgs field and the electroweak force results in mass being attributed to the intermediate vector bosons. The Higgs field also has a carrier particle of its own, the *Higgs boson*. With these extra assumptions, the electroweak theory seemed complete, and experimental physicists began their search for all four bosons.

The intermediate vector bosons were discovered experimentally two decades later in 1983 at the Super Proton Synchrotron in the European Organization for Nuclear Research (CERN), Switzerland, and the Higgs boson after half a century in 2012 at CERN's Large Hadron Collider. The latter is shown in Fig. 9.1. The huge resources required for experiments in modern particle physics is immediately apparent.

Fig. 9.1 The Large Hadron Collider (LHC) in CERN, the instrument used for the discovery of the Higgs Boson. Image courtesy of CERN (Image released by CERN under Creative Commons (2013) https://home.cern/news/news/cern/cern-releases-photos-under-creative-commons-licence (accessed 2020/06/3))

If the unification of the weak and electromagnetic interactions into the electroweak interaction represented the first step towards a Theory of Everything, it remains unfortunately the only step. So far, no unification has been achieved with the strong interaction. The weakest of the forces, gravity, also remains very much on the outer. As we have seen in Chap. 7, the theory of relativity interprets gravity as arising from a distortion of the geometry of space-time, and a quantum theory of gravity, analogous to QED, has not yet been found.

9.4   Sorting the Particle Zoo

Now that we have gained an insight into the role played by the four forces in the subatomic domain, it is time to return to the Particle Zoo, with a view to sorting out the exhibits that industrious experimentalists were rapidly discovering and attempting to classify. The advent of powerful new accelerators in the 20th century provided physicists with access to particles with energies that had hitherto been unattainable. So many "fundamental" particles turned up in experiments that it was soon realised that not all could be truly fundamental; many of them were probably constructed from other smaller units.

This suspicion was strengthened when the proton was examined more closely. We saw in Chap. 8 how electrons behave like tiny magnets when placed in a magnetic field. The strength of this behaviour is determined by the electron's magnetic moment, and its prediction to astonishing accuracy is one of the triumphs of QED. However, when, in the nineteen-thirties, measurements were made of the magnetic moment of the proton, the theoretical and experimental results were in mutual disagreement by hundreds of percent. This was soon recognised as evidence that the proton has internal structure of its own. The proton is *not* a fundamental particle, and like Alice in Nanoland, we must go to an even smaller domain, to search for the most basic constituents of nature.

The reader can probably guess one path physicists might take to further progress: follow the example of Rutherford in the atomic domain, and smash two particles together to see what eventuates. A high energy analogue of the Rutherford experiment described in Chap. 8 was carried out at the Stanford Linear Accelerator (SLAC) in California in 1970. Protons were bombarded with very high energy electrons[5]. The distribution of the electrons after the

---

[5]The energy was equal to that produced by accelerating the electrons through an electrical potential difference of 4.5 to 20.5 billion volts.

collision indicated the presence of internal structure in the proton [2]. There could be no doubt now: the proton definitely *was not* a fundamental particle.

Trying to get some sort of order into the assemblage of particles that had turned up in their collision experiments became a priority for particle physicists in the last half of the 20th century. We will not follow the history too closely, but rather attempt an overview of the currently accepted situation, and highlight any outstanding problems.

We have already indicated that particles can be divided into two groups, fermions and bosons, depending on their intrinsic spins. This division is important, because fermions satisfy the Pauli Exclusion Principle (see Chap. 8), while bosons do not.

However, there are other classifications that are important in attempting to sort out our zoo. Just as a visitor to a normal zoo might expect the big cats (lions, tigers, etc.) to be in a different area from the bears and insects, so physicists have arranged the particles into *leptons* and *hadrons*. The hadrons are themselves subdivided into *baryons* and *mesons*. We will treat these various groups separately, and see what their differences are.

Leptons

Leptons are particles that interact only via the electroweak interaction, and not the strong interaction. (We will not consider gravity further in our discussion of particles.) The name comes from the Greek root *leptos*, meaning "slight". This name is now somewhat inappropriate, as since the name was first adopted, leptons more massive than protons have been discovered, but no one has yet succumbed to the urge to change the name.

Leptons are truly elementary particles, having no internal structure. According to the Standard Model, all leptons are point particles; their volume is zero. What this means exactly is complicated by the Heisenberg Uncertainty Principle, which requires the particle wavepacket to occupy a non-zero volume. Physicists discuss the intrinsic "size" of a particle, i.e. the size of its internal structure, rather than the size of its wavepacket. Leptons have no internal structure, so their "size" is zero. Experimental evidence shows the electron to be smaller than 10^{-18} m, or at least 1000 times smaller than a proton.

Leptons come in three *generations*, each containing two *flavours*, so that we have six flavours in total. (We must now make our acquaintance with the somewhat flowery terminology of particle physics, where words do not have the same meaning as they do in everyday life. Particle physicists have adopted the attitude of Humpty Dumpty: *"When I use a word," Humpty Dumpty said*

Table 9.1 Table of Leptons

Generation	Name	Symbol	Antiparticle	Charge[a]	Mass[b]
1	Electron	e^-	e^+	−1	1.0
1	Electron neutrino	ν_e	$\bar{\nu}_e$	0	Small, but non-zero
2	Muon	μ^-	μ^+	−1	206.8
2	Muon neutrino	ν_μ	$\bar{\nu}_\mu$	0	Small, but non-zero
3	Tau	τ^-	τ^+	−1	3477.5
3	Tau neutrino	ν_τ	$\bar{\nu}_\tau$	0	Small, but non-zero

[a]Charges are expressed relative to the magnitude of the electronic charge
[b]Masses are expressed relative to the electron mass

in rather a scornful tone, "it means just what I choose it to mean — neither more nor less" [3].) A flavour is akin to the *species* of an animal in a normal zoo, and has nothing to do with our sense of taste.

In the Standard Model, Leptons are arranged as in Table 9.1.

As can be seen, Pauli's chimerical particle, the neutrino, now comes in three flavours. In the Standard Model, the neutrinos are massless. However, this is no longer believed to be the reality. A thirty-year old puzzle was why the number of neutrinos arriving on the earth from solar nuclear reactions was only about one third that expected from theory. The solution, which earned Takaaki Kajita of Japan and Arthur B. McDonald from Canada the 2015 Nobel Prize in physics, is that the missing neutrinos escape detection by changing flavours on their journey to the earth. This ability to change flavours requires the neutrinos to have a non-zero mass, however small that may be. These observations induce a sense of humility, as they reveal that the Standard Model, despite its great success, cannot be the complete theory of the fundamental constituents of the universe.

To distinguish leptons from other particles, physicists introduced a new quantity, called the *lepton number*. All the leptons in the above table have a lepton number of +1, and all the anti-leptons −1. Particles that are not leptons (e.g. neutrons, protons, etc.) have a lepton number of zero. Physicists believe that the lepton number is conserved during particle interactions (i.e. the sum of the lepton numbers before the interaction is equal to the sum of the lepton numbers after the interaction.)

For instance, consider the decay of a neutron into a proton, electron and antineutrino that we discussed earlier:

$$n^0 \rightarrow p^{+1} + e^{-1} + \bar{\nu}^0$$

The superscripts give the electric charges of the individual particles, and we see that the sum of the charges on the right-hand side is equal to zero, which is the same as the charge of the neutron on the left-hand side. Similarly, the lepton numbers, 0, 1, and −1 respectively, of the three particles on the right-hand side, sum to zero, which is the lepton number of the neutron. Both electric charge and lepton number are therefore conserved in this process.

The ability of a neutrino to change its flavour has led to the question being asked: is a neutrino its own antiparticle? (This is not as absurd as it may sound. Photons and gluons—to be encountered soon—are their own antiparticles.) In this case, lepton number may not be conserved in particle decays.

A final question springs to mind when surveying the table of leptons above: why stop with three generations? (This is the question Alice posed to Dr Quantum in our opening story.) How do we know that there are not more, heavier leptons, waiting in the wings to be discovered in the future, as our accelerators reach to ever higher energies? The same question can be asked when we come soon to examine quarks. The answer is the same in both cases: we don't know. However, fourth and higher generations are considered unlikely. Their presence would result in slight changes to predictions of electroweak theory that are not observed. Also, any extra fermions would interact with the Higgs boson, modifying its properties so that it would not have been detected. The observation of the Higgs boson, with the properties it has, is evidence that there are only three generations. Statistical analyses at CERN and the Humboldt University of Berlin exclude the presence of a fourth generation with a 99.99999% probability [4].

Hadrons

It is now time to consider the other large family of fundamental particles, those interacting by means of the strong interaction[6]. They are known as *hadrons*, and as we have seen, have two main subgroups, called *baryons* and *mesons*. Unlike leptons, hadrons are not truly elementary, in the sense that they are themselves composed of smaller particles called *quarks*.

The name "quark" was coined by physicist, Murray Gell-mann, when he noticed the sentence: "*Three quarks for Muster Mark*" in James Joyce's book, *Finnegan's Wake*. At the time he believed that there were three quarks occurring in nature. Now it is known that there are six, and they have been given

---

[6]They may also interact via the electroweak interaction, but not exclusively so, as is the case with leptons.

the rather uninspiring names: up (u), down (d), charm (c), strange (s), top (t) and bottom (b). For a while, "truth (t)" and "beauty (b)" were floated as possible names for the last two, but banality ultimately prevailed. Most physicists pronounce "quark" to rhyme with "Mark"; Gell-mann pronounced it "kwork".

At first, considerable effort was expended trying to detect free quarks in cosmic rays and accelerator data. Quarks have an electric charge of 1/3 or 2/3 that of the electron, which should make them distinctive. Despite a few false alarms [5], it is now generally accepted that quarks are confined within the hadron, and cannot be seen on their own. We shall discuss this idea further shortly.

In a conventional zoo, it is customary to put a label on the animals' cages bearing the names of the beasts within. For instance, the tigers' cage would have the tag: *Panthera tigris tigris* if it contains a Bengal tiger, or *Panthera tigris sumatrae* if there is a Sumatran tiger lurking somewhere back in the shadows. The three Latin names are the genus, species and sub-species of that strain of tiger. In our particle zoo, *quantum numbers* serve the same purpose, i.e. they enable us to differentiate between particles. (We have already encountered the lepton number, which is one example of a quantum number.) Unlike the Latin scientific names of animals, however, they also limit the interactions that particles can engage in, via various conservation laws.

Very early in nuclear physics, in 1937, a quantum number called *isospin* was introduced by Eugene Wigner, in analogy with ordinary spin, to distinguish protons from neutrons. Particles that are affected equally by the strong force, with similar properties, apart from their charge (e.g. neutrons and protons), are considered to be different states of the same particle. They are said to have different isospin. Later, physicists introduced the additional quantum numbers of baryon number, charm, strangeness, topness, and bottomness to distinguish the newer members of the particle zoo. Not all physicists had Gell-man's erudite knowledge of literature to draw on when choosing names. A table of the six quarks and their associated quantum numbers is included in Appendix 9.2.

We are now almost in a position to see how the individual hadrons are put together. However, simply forming all possible combinations of the known quarks would result in far more hadrons than are actually observed. Clearly there is another selection process at work here. This involves what has been called *colour*, although it has nothing to do with the visual colours that we know so well.

Just as the electromagnetic (or more correctly, the electroweak) force binds the electrons to the atomic nucleus, colour (or the *colour force*) binds the quarks together inside a baryon, e.g. inside a proton or neutron. The mediating carriers for this force are the *gluons*. They are called *gluons* because they are the source of the glue binding the quarks together.

So how does all this impact on the strong interaction which, as we saw in Sect. 9.3, binds the protons and neutrons together inside a nucleus, and is responsible for the energy of the atomic bomb? An analogy can be drawn with Van der Waals forces in chemistry. As two atoms approach together, their electron clouds repel each other. Hence, although the atoms are electrically neutral, their negatively charged electrons and positively charged nuclei do not remain coincident. As a consequence, the atoms experience a weak attractive force, which is a *residual interaction* left over from the imperfect cancellation of the negative and positive fields of the electrons and nuclei. As the atoms come closer together, their electron clouds begin to overlap, and the attractive force turns into a strong repulsive one. Van der Waals forces play an important part in some physical processes. They have been offered as an explanation for the force enabling a gecko to hang from a sheer glass wall [6] (see Fig. 9.2), although this explanation has been challenged.

In the Standard Model of Particle Physics, remnants of the colour force extend outside the boundaries of the neutrons and protons. It is this residual force that produces the *strong nuclear interaction* that is responsible for most of nuclear physics, including the destructive power of nuclear weapons.

Fig. 9.2 Giant leaf-tail gecko, *Uroplatus fimbriatus*, clinging to glass. Image from Tom Vickers (Wikimedia Commons, public domain. https://commons.wikimedia.org/wiki/File:Uroplatus_fimbriatus_(3).jpg (accessed 2020/6/8))

Within the hadron, the colour force is unlike the other forces, and does not drop off with distance. This property is responsible for the confinement of the quarks within the hadron, and is the reason no free quarks are observed in laboratory experiments. Indeed, the colour force, produced by the exchange of gluons, is so strong that quark-antiquark pair production will occur before the quarks can be separated. However, at short separation distances the colour force is weak, and the quarks can move freely, until they start to get too far apart. Then the colour force kicks in and brings them back together, rather like a sheep dog rounding up lambs that have strayed too far from the flock.

Colour is the analogue in the colour force to charge in the electromagnetic force. However, whereas two values were needed for the electric charge, denoted by convention "positive" and "negative", three values are needed for the colour. These three values of the colour "charge" have been given the names of the three primary colours: red, green and blue. Clearly the colour charges have nothing to do with visible light, or the perception of colours. They are just another example of the Humpty-Dumpty influence that we mentioned earlier.

Without the existence of colour, the Pauli Exclusion Principle would not allow, for instance, two up quarks to be present inside a hadron. However, if each up quark is of a different colour, they are non-identical, and therefore can be present together.

Having got our heads around the basic concepts of the Standard Model, we are now at last in a position to see how hadrons are put together. There are some limitations on how these particles are constructed. First, only "*colourless*" hadrons exist in nature. These may be baryons comprising three quarks of the three different colours (in the same manner that white light is produced by combining the three primary colours of red, green and blue light), or mesons made from a quark-antiquark pair. Antiquarks have an *anticolour*, so that a quark-antiquark pair is colourless. In addition, as no particles with fractional electronic charges exist in nature, baryons must have an electric charge that is an integer multiple (disregarding the sign) of the electronic charge. As they are fermions, they must also have half-odd integer spin.

Now would seem to be an appropriate time to introduce a table of hadrons, showing how they are all constructed from their constituent quarks. However, thanks to the diligence of experimental high-energy physicists, there are literally hundreds of them, and studying such a list would, for the non-specialist, be about as entertaining as reading the Greater London Telephone Directory, or perhaps even Whitehead and Russell's *Principia* (see Chap. 3). For illustrative purposes, we discuss here only the proton and the pion, as examples of a baryon and a meson.

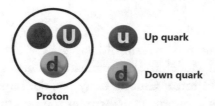

Fig. 9.3 The quark structure of the proton

In the Standard Model, the proton (see Fig. 9.3) is comprised of two up quarks and one down quark. The sum of the electric charges of the three quarks is +1 (2/3 +2/3 −1/3) and the sum of the baryon numbers is 1 (1/3 + 1/3 + 1/3). The spins of the three quarks can be oriented to give a spin of +1/2 for the proton, and the same is possible for the isospin. (The isospin of a proton is +1/2 and that of a neutron −1/2.) The presence of the three differently coloured quarks ensures that the proton is colourless.

The astute reader may have noticed, by referring to the Table of Quarks in Appendix 9.2, that the sum of masses of the three individual quarks comprising the proton amounts to no more than 1% of the proton mass. The remaining mass comes purely from the motion and confinement of quarks and gluons [7], in accord with the equivalence of mass and energy that we discussed in Chap. 7.

We turn now to the pion, or pi meson, as an example of that other class of hadrons, the mesons. There are three pions, π^+, π^-, and π^0, with charges of +1, −1 and 0 respectively. The structure of the π^+ pion is shown in Fig. 9.4. As we can see, there is a quark and an antiquark present in the pion. An antiquark can take one of three anticolours, called antired, antigreen and antiblue. To continue the analogy with the familiar colours of light, these are represented by cyan, magenta and yellow respectively.

The remaining members of the particle zoo can be explained in a similar fashion from conglomerations of quarks, and the observance of appropriate conservation laws for charges and quantum numbers.

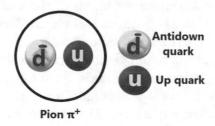

Fig. 9.4 The quark structure of the π^+ meson

Gluons

Before quitting our zoo, we should say a little about the final components of the Standard Model, the unfortunately named gluons, carriers of the colour (or interquark) force. Gluons carry out the same role for the colour force between quarks as photons perform for the electromagnetic interaction between electrons. However, they are also carriers of the colour charge, unlike photons, which do not carry the electromagnetic charge. They are massless, like the photon, but unlike the photon, they are never observed outside the confines of the hadron. They have spins of 1 unit, and are therefore bosons.

Gluons carry both a colour and an anticolour. This gives a possible nine combinations. An exchange of a gluon between quarks converts a quark from one colour to another. The analogy of this process with photon exchange has led to the name Quantum Chromodynamics (QCD) being coined for this aspect of the Standard Model.

9.5   Beyond the Standard Model

In the last chapter we have made much of the incredible agreement between QED predictions and experimental measurements for physics involving the electron-electron interaction. Naturally physicists hoped to achieve the same precision with the strong force analogue to QED, i.e. QCD. As we shall see, they have been only partially successful. Various reasons have been put forward for this, and suggestions made to extend the Standard Model. As this is currently a highly active research field, we should bear in mind that the situation can change very quickly.

In both QED and QCD, the force is produced by the exchange of carrier particles. This is illustrated by the Feynman diagrams below in Fig. 9.5. These are the lowest order diagrams for three processes. As we mentioned in Chap. 8, there are higher order diagrams, and all possible diagrams should, in principle, be included in the theoretical calculations.

In (a) the decay of a neutron by means of the weak interaction into a proton, an electron and an antineutrino is displayed. The process is mediated by the exchange of an Intermediate Vector Boson.

In (b) an example of the colour force in action is shown. A green quark and a blue quark swap colours by the exchange of a green-antiblue gluon.

The last example in (c) is the scattering of a neutron and a proton inside the nucleus by the exchange of a pion. This is the process that was proposed by Yukawa when he predicted the existence of the meson to explain the strong nuclear interaction between nucleons inside the nucleus. Now, if the strong

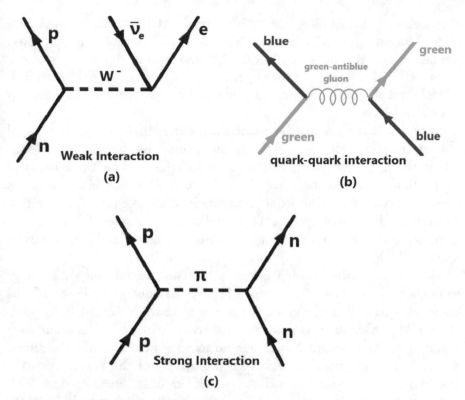

Fig 9.5 Feynman diagrams for the exchange of carrier particles in the various nuclear interactions

interaction is interpreted as a residual interaction left over from the inter-quark colour force inside the nucleons, there is some doubt over the role, if any, played by pion exchange.

If physicists hoped for a level of agreement in QCD comparable to that of QED for the electron then they were disappointed. We have seen in Chap. 8 how accurate the QED prediction of the anomalous magnetic moment of an electron is. A similar calculation for the muon requires the inclusion of Feynman diagrams related to the weak interaction. In this case, the predicted muon anomalous magnetic moment differs from the measured value [8] by 3.5 standard deviations. (From probability theory, one expects 99.7% of measurements to lie within 3 standard deviations of the correct result.) However, the agreement is still to the eighth decimal place in the actual magnetic moment[7]. On the other hand, composite particles such as the neutron and proton have huge anomalous magnetic moments.

---

[7] Predicted α_μ =0.001 165 918 04(51), measured α_μ =0.001 165 920 9(6).

Another substantial disagreement between the Standard Model and experimental measurement is in the model's prediction of the proton radius, where the discrepancy is enormous[8]. In this case, and also for the muon magnetic moment, physicists are reluctant to attribute the discrepancies to the Standard Model until all possible sources of error in the experimental measurements have been eliminated.

As well as the discrepancies mentioned above, there are some physical phenomena which are left completely unexplained by the Standard Model. For instance, gravity remains steadfast in its refusal to be accommodated in the same scheme as the other physical forces. Since Einstein, gravity has been interpreted as arising from curvature in the space-time fabric within which the other forces operate. This fundamental incompatibility between quantum mechanics and relativity is one of the major embarrassments of modern physics.

More recently, the growing evidence for dark matter and dark energy throughout the universe (see Chap. 11) proposes another challenge for the Standard Model. It might be expected that a theory of fundamental particles would provide an explanation for these two phenomena, but as currently constituted, the Standard Model has no relevant mechanism. In addition, a particle that has been around since the middle of the last century, the neutrino, has been found to oscillate between three different flavours. Such a characteristic requires the presence of mass, but in the Standard Model the neutrino is massless. This disagreement is not yet resolved.

Much has been made of the discovery of the Higgs boson as a triumph for the Standard Model. There are, however, other hadrons that are predicted at very high energies by the Standard Model, and have not yet been observed. In addition, "glueballs", which are particles comprised solely from two or three gluons, but no other hadrons, are predicted. Their detection would represent a major coup for the Standard Model.

Other forms of "exotic" particles have indeed been discovered, including pentaquarks and tetraquarks, composed of five and four individual quarks respectively. Recently it was announced by CERN that a tetraquark comprised of four charm quarks had been discovered [9]. It is a good candidate to test theories on whether the four quarks are tightly bound, or arranged to form two mesons, which are stuck together loosely in analogy to a chemical molecule.

Even though the Standard Model has had great success in establishing an order in the chaos of the particle zoo, it has fundamental limitations. We

---

[8]About seven standard deviations, corresponding to a 1 in 390 billion chance that the model and measurement are in agreement.

saw in Chap. 8 how Feynman lamented that the Fine Structure Constant cannot be calculated by physicists, but must be measured by experiment and inserted into the equations of QED. Philosophically this is quite unsatisfactory because even a comparatively small change in the value of this constant would result in a very different universe that would not support life as we know it.

However, the Standard Model has nineteen independent constants which have to be inserted into it arbitrarily. Examples of these constants are the masses of the elementary particles and of the Higgs boson, as well as constants, analogous to the Fine Structure Constant, which govern the strength of the various interactions. Such a degree of arbitrariness is clearly not appropriate, and physicists are continually searching for other approaches.

Two avenues that have been explored to extend the Standard Model are a Grand Unified Theory (GUT) and Supersymmetry. We have discussed these approaches briefly in Appendix 9.3. Both of these theories predict the existence of new particles. However, unfortunately these new particles are so massive that it is beyond the capabilities of existing particle colliders to produce them.

String Theory is another contender that has received a lot of attention in the research literature and in the popular media. It was first developed in the nineteen-sixties as a possible explanation of the strong nuclear force. It fell from favour as Quantum Chromodynamics became popular, but has found another life as a candidate for a Theory of Everything (TOE), perhaps the Holy Grail of physics (see Chap. 4). The attraction of String Theory is that it treats gravity on the same footing as the other forces, and can therefore be regarded as a theory of quantum gravity.

As the name might indicate, in String Theory "zero-dimensional" point-like particles are replaced by one dimensional "strings". These may be either lengths of string, or loops of string. The properties of the particle (mass, charge, etc.) are determined by the vibrational states of the string. One of these vibrational states corresponds to the *graviton*, the long sought-after carrier particle of the gravitational field. Originally String Theory only included bosons, but has now been extended into *Superstring Theory*, to also include fermions.

Five different versions of String Theory were developed over the years until it was found that they were all variants of one overriding single theory, called *M-theory*. M-theory is formulated in eleven dimensions, which represents an improvement on the original bosonic string theory, which required twenty-six dimensions.

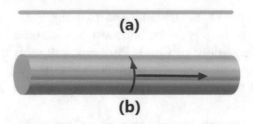

(a)

(b)

Fig 9.6 **a** A length of hose appears one dimensional when viewed from a distance. **b** When viewed close up, the two dimensional nature of the hose surface becomes apparent. The transverse dimension is "closed" because if one travels in this direction, remaining on the surface, one returns to the starting point

As the "familiar" space-time of Einstein is only four-dimensional, the question arises: what has happened to the missing seven dimensions? One explanation is *compactification*, where the extra dimensions are assumed to close up on themselves. Imagine a length of cylindrical hose (see Fig. 9.6).

From a distance the hose looks one-dimensional, but as one gets closer, it becomes clear that the surface is two dimensional. One dimension lies along the length of the hose, and the second runs transversely around the perimeter. The second dimension is closed, because if an ant on the surface were to travel around the perimeter of the hose, it would finish up back at its starting point.

In the case of String Theory, the extra seven dimensions are thought to be closed into very tight circles. They are then unobserved in the same manner that the transverse dimension of the hose is not noticeable at large distances.

String Theory has been applied to many of the outstanding problems of physics, covering the full scope from fundamental particles to cosmic inflation (see Chap. 11). However, it attracts criticism because it can "explain" too much. As currently understood, it can describe around 10^{500} different universes. From this inconceivably high number of possibilities, there is surely one to describe our universe, no matter what it happens to be. If you want to play the lead role in the Royal Shakespeare Company's next production of Hamlet, there is already a universe waiting for you in String Theory where that can happen. All experimental attempts to verify the theory must remain unconvincing if, no matter what result is obtained from the measurement, it can be explained by selecting the appropriate universe.

As we can see from the above arguments, although the Standard Model has achieved much in providing a framework for the interpretation of the particle zoo, it also has its limitations. Many physicists believe it is an interim step along the way to an understanding that will only be achieved when a better, less arbitrary, theory is discovered.

Since the time of Galileo, experiment and theory have each contributed equally to our understanding of the physics of our world. Sometimes experimentalists have discovered new phenomena that have challenged theorists to produce new theories to explain them, and sometimes theorists have told experimenters which experiments are critical to distinguish between competing theories. In the main, a happy partnership has existed between two groups of physicists with very different skill sets.

In the field of Fundamental Particle Physics, however, the cost of the accelerators needed to reach the energy regions now of interest to theorists has become so high that these machines strain the budgets of even the wealthiest of nations. There are so many other competing priorities for scarce resources that it is possible that the current generation of particle colliders is the last we shall see for the foreseeable future. If that is the case, it falls to theorists to find alternative ways to test their theories. This in itself represents a new challenge for the next generation of particle physicists.

References

1. Gardner M (1964) The ambidextrous universe: left, right and the fall of parity. Penguin Books
2. SLAC News, vol 1, no 1, Feb 26, 1970
3. Carroll L (1896) Through the looking glass
4. Eberhardt O et al (2012) Impact of a Higgs Boson at a mass of 126 GeV on the standard model with three and four Fermion generations. Phys Rev Lett 109(24):1802
5. McCusker CBA (1983) An estimate of the flux of free quarks in high energy cosmic radiation. Aust J Phys 36:717–23. https://www.publish.csiro.au/PH/pdf/PH830717. Accessed 5 June 2020
6. Autumn K et al (2002) Evidence for van der Waals adhesion in gecko setae. Proc Natl Acad Sci 99(19):12252–6
7. Walker-Loud A, https://physics.aps.org/articles/v11/118#c1. Accessed 31 July 2020
8. Patrignani C et al (2016) Particle data group. Rev Particle Phys (Chin Phys) C40, p 32
9. https://theconversation.com/cern-physicists-report-the-discovery-of-unique-new-particle-142315. Accessed 31 July 2020

10

How the World Began

10.1 Cornetti for Breakfast

Giuseppe was a simple market gardener, as was his father before him, and his grandfather before *him*, and so on. Giuseppe was not sure how long his family had been growing fruit and vegetables for the local market at *Castellina in Chianti*, Italy, but it had been for a very long time. Without doubt, this gave him some rights, or so he believed. He smiled to himself at the frustrated long line of vehicles behind his tiny three-wheeled *Ape* as, laden with vegetables, he made his way into town each Saturday along the narrow Tuscan roads.

Horticulture had always been Giuseppe's life. Even at school, he had interpreted every lesson to see how it might be used to improve the vegetable gardens that one day would be his to run. This interest in advancing the family business had all started, strangely enough, when his class teacher explained some basic concepts of geometry, such as circles and spheres.

"It is a remarkable thing", she used to tell the class, "that the circumference of a circle is proportional to its radius, while its surface is proportional to the square of the radius. That means that if you double the radius, the circumference only doubles, but the surface becomes four times as big." She paused to let the children take in the implications of her remarks, then reached into a drawer and held up an apple that she had brought in for

© The Author(s), under exclusive license
to Springer Nature Switzerland AG 2021
R. Barrett and P. P. Delsanto, *Don't Be Afraid of Physics*,
https://doi.org/10.1007/978-3-030-63409-4_10

a snack. "But in a sphere," she continued, "the volume becomes eight times larger."

This lesson was not lost on the young Giuseppe, and at the market the next Saturday, he peered at a tray of strawberries, some as large as a tomato, and at tomatoes the size of a cherry, and an idea began to form. "If I can produce larger and larger fruits and vegetables, it will be much quicker to peel them, since the volume grows much faster than the surface to be peeled. And it is not only a question of time, but also of economy, since by peeling less, there will be less wastage."

So in his spare time, Giuseppe began to carry out experiments in a little-used greenhouse at the rear of the farm. He was unsuccessful however, and eventually gave up. It was not until years later, when he finally inherited the farm, and was walking its perimeter, that he saw the old greenhouse and recalled his investigations. One morning, while he was spreading a thick layer of strawberry jam on his breakfast cornetto, Giuseppe let his mind drift back to those earlier experiments. He quaffed his macchiato in one gulp and reached for another two cornetti, and that was the precise moment that the solution came to him. He couldn't wait to try out this new technique to control plant growth hormones.

This time Giuseppe was successful. He refused to divulge his secret to other growers, and slowly his products became sought after, and soon he could hardly keep up with demand. All his fruit and vegetables were grown to fixed weights, so that they could be sold without the need to weigh them, and with uniformly high quality. In addition, they were 100% "organic", and not in the least genetically modified.

One morning a good friend of his, the cook of a local Greek restaurant, came to buy two 5 kg eggplants for preparing his renowned moussakas. He knew that their weight was reliable. Nevertheless, he insisted on their being weighed, so he could express his wonder, and repeat his favourite joke: "Next you'll have to grow them cubic to better suit the shape of my baking pans And also grow them some legs, so they can run down the hill to my restaurant, saving my arthritic knees." Giuseppe laughed at the old gag, and put the eggplants on his scales. To his great surprise, he found that they weighed 5.1 kg each, 100 g above their nominal weight.

Later, back home again, Giuseppe was still disturbed by the incident, and he decided to weigh his 5 kg fruits and vegetables. All of them were now 5.2 kg. He was very confused, but cheered up a little when he realised that at least now his profits would increase. Something, however, was definitely not right.

The next morning, as soon as he got out of bed, Giuseppe felt slightly uncomfortable. He seemed to require more of an effort to move. His stomach rumbled, but that was usual before breakfast. On an impulse, he walked into the bathroom and stepped onto the scales. *Che Diavolo*! He had gained 2 kg since yesterday morning.

He was really puzzled now, as he entered the kitchen and began to prepare his morning coffee, and warm up his cornetti. He switched on the radio to his usual news station, and caught the end of an item on a certain Mr Newton, and how experts believed his constant might have suddenly begun to change.

"Stupido," he proclaimed aloud. "If constants aren't constants, then I am not Giuseppe, and these eggplants are pineapples." Annoyed, he changed the station, but there was more nonsense here as well. They also were talking about changing weights, but this time the culprit was Mr Higgs and some bosons he'd found. "What the hell are bosons, I wonder?" he muttered.

Giuseppe returned to the bathroom and stepped onto the scales again. His weight had already increased by one more kilo. It seemed that a force greater than anything he could muster was determined to make him gain weight. Well, so be it. At least there were compensations. "When there's nothing to be done, it's better not to resist" his father had once advised him. He returned to the breakfast table, and placed two more large cornetti on his plate. His saliva was already flowing as he covered them in jam and cream, and took the first massive bite of his breakfast. Whatever these bosons were, they clearly wanted him heavier, and who was he to argue?

* * *

As we shall see later in the next two Chapters, whether constants remain constant is an important question in the study of cosmology, and one that it is very difficult to answer.

10.2 The Very Beginning

In Chap. 2, we explored the possible origins of human intelligence and the evolution of logic. As these aptitudes evolved, early humans applied their new skills to make sense of the world in which they lived. Their curiosity extended to the search for an explanation of the origin of the universe, a study now known as *cosmogony*. They also pondered about their own origin, as well as making hypotheses for what might lay ahead. Thus were born a variety of religions in the different centres of human population.

Ancient mythology is a fascinating subject in its own right. However, what interests us here is the first appearance of ideas that are still part of modern cosmology. The concept of time was personified in pre-Socratic philosophy in the figure of Chronos, or *Father Time*, wielding his harvesting scythe, and wreaking havoc on all about. Was this a forerunner of *the arrow of time* and the Second Law of Thermodynamics (see Chap. 2), which states that the world moves from a state of order to disorder? The Second Law can perhaps also be recognised in Greek poet Hesiod's five ages of the world, going from the best (golden) to the worst (iron), and in the four epochs of the great Indian texts, *Vedas* and *Puranas*.

In both the Chinese and the Hindu traditions, the evolution of the universe is conceived as a cyclic process, with the Hindu cosmology even quantifying the duration of each cycle in terms of billions of years. At the end of each cycle the universe is destroyed by fire, and then after an interlude, it is created again. As we shall see, this cyclic nature appears also in some modern conjectures about the cosmos, extending Einstein's General Theory of Relativity beyond the initial (Big Bang) and possibly final singularity (Big Crunch).

Interesting as these ancient insights are, we cannot attach too much significance to them. There are a limited number of options for the evolution of the universe: static, cyclic, expanding and shrinking, so it is not surprising that primordial mythologies have suggested some of them. The important thing is that humans, so early in their history, pursued an explanation of the nature of their world.

The birth of what we regard as modern cosmogony had to await accurate observations of the heavens. These suggested that the earth was a sphere, and that the skies revolved around it diurnally. The use of the plural here is necessary because the sun and the moon revolved on a different layer (sky) from the stars. In addition, the observations also revealed a small number (five) of *wandering stars,* or *planets.*

The movements of the planets were not easy to understand, since sometimes they even went in the *wrong direction* with respect to the other stars. At the beginning of the second century CE, Claudius Ptolemy, an astronomer of Alexandria, elaborated a geocentric description of the universe, in which an articulated machinery explained all of the apparent anomalies in the motions of the celestial bodies.

This mechanism is explained in Fig. 10.1. The earth lies off-centre from a circle, known as a *deferent*. The planet revolves around an *epicycle*, the centre of which revolves around the deferent. Another position, known as the *equant* lies opposed to the earth on the opposite side of the centre of the deferent. The significance of the equant is that to a hypothetical observer located there, the centre of the planet's epicycle moves at a constant angular speed around the deferent. The presumption that all planetary orbits are circular (and all celestial bodies are perfectly spherical) was an article of faith for Aristotle and his followers, and the equant helped to maintain this illusion.

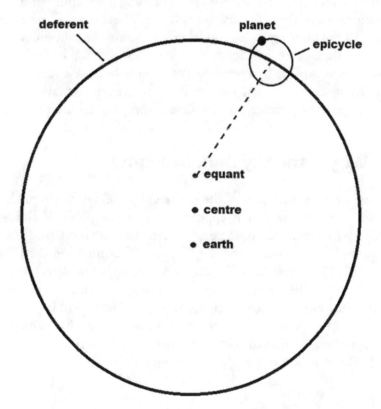

Fig. 10.1 Example of the Ptolemaic model to describe a planetary orbit, showing the use of a deferent and epicycle

This explanation of the motion of stars, sun, moon and planets held until the Sixteenth Century, when Copernicus revived an earlier, and forgotten, heliocentric model by Aristarchus. The troubles accompanying the Reformation, and conflict between different political powers (including the Roman Church), made the Copernican revolution particularly dramatic, as we know from the history of Galileo, who was forced to recant his belief in the Copernican model.

From the Seventeenth century, modern science became pervasive in all areas, including astronomy and cosmology. Telescopes allowed much more powerful observations of the sky. Newton's theory of universal gravitation shed new light on celestial mechanics, and it was realised that planetary orbits were elliptical, not circular, removing the need for epicycles. At the end of the Nineteenth century, the description and interpretation of the universe in scientific terms began.

However, even today the Ptolemaic model still finds an application. The projectors for modern planetariums are built using gears and motors to carry out the functions of the deferents and epicycles of Ptolemy. Their success is a testament to the accuracy of his model at predicting the motion of heavenly bodies, even though we now know that it has no physical basis. The planetariums perhaps provide us a warning that simply because a model describes a set of observations, it is not necessarily a correct representation of nature. Figure 10.2 shows a photograph of a Zeiss Klein planetarium projector.

10.3   Why is the Sky Dark at Night?

Let us begin our explorations of the universe by asking the simple question above, which on the face of it appears naïve to the point of childishness. Certainly the reaction of most people when first encountering this question is to reply somewhat scornfully: "because the sun is on the other side of the earth". The retort is usually accompanied by a few derogatory asides. However, what on the surface appears an inane query contains within it some deeply puzzling aspects when we rigorously apply the tool of logic, which we have discussed in Chap. 3. So much so, that the question has been given the name *Olbers' Paradox* after the German astronomer Heinrich Wilhelm Olbers (1758–1840), who was one of the first to raise it.[1]

---

[1] Actually, this paradox, in one form or another, had been considered since at least a couple of centuries before Olbers, who formulated it in 1823.

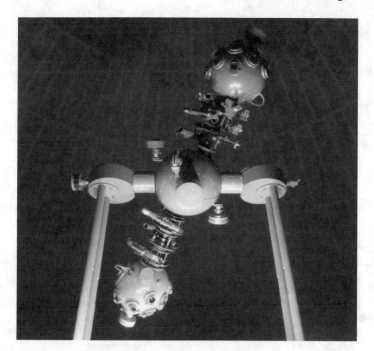

Fig. 10.2 Zeiss Klein Planetarium 2P Projector. Image courtesy of Eryn Blaireová (https://commons.wikimedia.org/wiki/File:Zeiss_Klein_Planetarium_2P.jpg   under Creative Commons Attribution Share Alike Licence)

To see the origin of the dilemma, we must first familiarise ourselves with the world view prevailing in the nineteenth century. The common representation of the Universe among astronomers (and scientists in general) at that time was of an infinite, substantially homogeneous, expanse of stars similar to the sun, with no location within the universe being more privileged than any other. Every object within the universe had its own evolutionary path, but anything dying was always replaced by something else being born, so that on the average the Universe would always appear to be the same, lasting forever, or at least indefinitely.

By this time Astronomers had accumulated data on the distance from the earth to the planets and nearby stars. On earth the measurement of length is a fairly straightforward process, requiring only the use of a tape measure, or some other suitable measuring device. However, astronomers are unable to run out a tape to the moon and stars. Today, the reflection of laser light and radar waves can be used for determining the distance to the moon and nearby planets. In the nineteenth century, astronomers obtained astronomical distances by using their telescopes to measure the parallax of the closer stars. Parallax occurs when nearby stars observed from the earth appear to

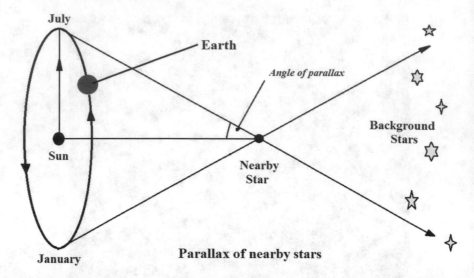

Fig. 10.3 The different positions of the earth in its orbit around the sun results in a parallax effect whereby a nearby star appears in a different position against the distant background of stars

move against the background of distant stars as the earth circles the sun (see Fig. 10.3).

A simple example of parallax can be observed by holding up an object at arm's length, and observing it against a distant background, shutting first one eye and then the other. The position of the object against the background will be different for each eye because of the separation of the two eyes. This effect becomes smaller as the object is moved farther away. Astronomers had verified that most visible stars (but not all) displayed zero parallax during the year, from which they deduced that these stars must lie at enormous distances from the earth (and the sun).

In the prevailing view at this time, there was no logical reason for any privileged place, or *centre*, of the universe. Why should the presence of stars end somewhere, leaving an infinite emptiness beyond? The universe must have no border and no centre.

What was true for space should also hold for time. There was no *special* moment here either; the universe had to be unchanging in time. In this case, the argument was more delicate and sensitive. If not to reject, at least to mark a complete separation from religion, no *origin* could be envisaged by science. A few centuries earlier, views such as these had led to Giordano Bruno, an Italian Dominican friar, philosopher, mathematician, poet, and astrologer, being charged with heresy and burned at the stake in 1600 CE. (Best-selling author, Morris West, has written a novel and a blank-verse stage play on the life of Giordano Bruno [1].) Astronomers in the 19th Century rejoiced in their hard-won emancipation and freedom of speech.

It is this conjecture of an infinite and uniform distribution of stars in the universe that gives rise to Olbers' Paradox. If you assume that there are infinitely many stars scattered everywhere and at all distances, all lines of sight from your eye terminate sooner or later on the surface of a star. Of course, the farther you go, the smaller the apparent size of the star. However, even if the apparent size is reduced to an infinitesimal point, the luminosity will always be that of the surface of a star. As a result, the sky should always be as bright as the surface of the sun.

When this argument is proposed, it is generally received with a shake of the head, and the comment: "but, you are assuming that the skies are perfectly empty and transparent. Aren't there dust clouds in space that cut off the light from stars behind them?" This is certainly true. The central region of our galaxy, the Milky Way, happens to lie in the southern skies, and as a consequence, the Milky Way in the Southern Hemisphere is brighter than in the Northern. Dust clouds are very noticeable against this prevailing brightness, especially a large one, known as the *Coalsack*, adjoining the iconic constellation, *Crux*, or the Southern Cross.

In Fig. 10.4, the Milky Way is shown as it appears in the southern skies, with the many dust clouds along its length clearly visible. The Coalsack is the separate dark patch in the Milky Way at the level of the top of the large telescope. It lies above and adjoining the two brightest stars of the Southern Cross. The Coalsack Dark Nebula is located at an approximate distance of 600 light years in the Constellation *Crux*. Its radius is 30–35 light years.

The distributed dust clouds, not the bright stars themselves, form the "constellations" of Southern Hemisphere cultures, e.g. the emu of various Australian Aboriginal peoples. The Coalsack forms the head and beak of an emu, the elongated dark strip above it is the neck, and the many dark clouds around the galactic centre form the body of the emu, which in this photograph is inverted.

These dust clouds clearly obscure the stars behind them, so the objection raised earlier would appear to have merit. However, it stumbles against the alleged eternity of the universe: if the present age of the cosmos is infinite, dust, radiation, stars, planets, everything, should have had time to reach thermal equilibrium. A dust grain absorbs radiation having the same temperature as the stars, warms up a bit, then re-emits the absorbed energy in the infrared (even far infrared) band of frequency (not visible to our eyes). A similar process occurs when a stone wall becomes warm from the afternoon sun, and continues to radiate warmth (aka infrared radiation) long after the

Fig. 10.4 The Milky Way and its associated dust clouds. This image was taken by the ESO Ultra High Definition Expedition team at the location of the ALMA antennas on the Chajnantor plateau, at 5000 m altitude in northern Chile. Image credit: ESO/B. Tafreshi (twanight.org) (Image courtesy of the European Southern Observatory/B, Tafreshi (twanight.org), from Creative Commons at https://www.eso.org/public/ima ges/uhd_img_2528_cc/)

sun has set. However, if the system is in thermal equilibrium, the dust will have had an eternity to reach the same temperature as the incoming radiation, and will emit at the same frequency as is incident upon it. Again, the sky should be as bright as the surface of a star.

Another possible objection is: "light takes time to travel from the source to our eyes, so the photons from the farthest stars simply have not yet arrived on earth". This remark implicitly assumes an origin of time, i.e. a beginning. If the universe is eternal, there will always be light arriving, no matter how far away the sources are located. After all, the light has forever to make the trip.

So we see that far from being naïve, the question: *"why is the sky dark at night?"* challenged the prevailing 19th Century view of the universe as homogeneous, eternal and infinite. (Any glance at the night sky will reveal that it is not really homogeneous, being full of constellations of stars, nebulae and dust clouds. However, we are considering here a much bigger picture, where these local concentrations are averaged out.) If we trust our logic, at least one of these three assumptions must be wrong, for our logic leads us to conclusions that are contradicted by observation. Despite all of the above objections, nineteenth century scientists did not appear to be overly troubled, and in the early times of Einstein, the stationary, uniform and infinite universe was the common wisdom, shared by Einstein himself.

In the next two Sections, we will see which of the 19th Century assumptions were erroneous, but we will also learn that the sky is indeed bright at night. It is simply that the "light" arriving from the depths of space has been red-shifted to frequencies far too low to be visible to our eyes, and must be detected by sophisticated equipment designed for that purpose. This discovery was one of the major achievements of 20th Century astronomy.

10.4 The Expanding Universe—Theoretical Framework

It is now time to advance to the Twentieth and Twenty-first Centuries, which saw an incredible expansion of our knowledge and interpretation of the world about us. This was brought about by the development of powerful new instrumentation, and by the revolutionary ideas of Quantum Mechanics and the Special and General Theories of Relativity. We have discussed these theories in Chaps. 5, 6 and 7 of this book, and we shall now see their importance in the understanding of the universe.

The driving force in the Cosmos is gravity. This may seem a little strange, because gravity is by far the weakest of the four fundamental forces that operate in nature. So what is going on here? Why is a force that has only a strength of about one billion trillion trillionth that of the weak nuclear force so important, both on earth (in holding us on the ground) and in outer space? The answer lies in the range of the forces, and the enormously large masses of the objects involved (compared with the masses of fundamental particles). The two nuclear forces have extremely short ranges and as a consequence do not extend beyond the radius of the nucleus.

Both the gravitational and electromagnetic forces are long range. However, the electromagnetic force can be either attractive or repulsive. Positive and negative electrical charges both exist in nature, and the force between similar charges (i.e. either both positive or both negative) is repulsive, and the force between dissimilar charges (i.e. one positive and one negative) is attractive. As the number of positive and negative charges are on the average equal and evenly distributed, the attractive and repulsive forces tend to cancel each other out. The gravitational force has no repulsive component, and so the forces between objects reinforce each other. As a result, when bodies are the size of stars and planets, the gravitational attraction between them is powerful enough to keep the planets in orbits about their suns, and to influence the creation and motion of galaxies.

One of the major achievements of Sir Isaac Newton was his Theory of Gravity. As we saw in Chap. 7, Einstein in his General Theory of Relativity, proposed a geometric interpretation, in which the curvature of space–time produces the effects of gravity. However, Einstein's theory is formulated with equations that are horrendously difficult to solve. Only a few special cases lend themselves to an analytical solution. One of these, which we encountered in Chap. 7, is the FLRW model of a universe filled uniformly with a dust cloud.

The aspect of the solution of this model relevant to our discussion here is that such a universe must either expand or contract: no steady state is allowed. The universe can be pictured to be analogous to the surface of an inflating (or deflating) balloon. The two dimensional surface of the balloon represents space (which in reality is actually three-dimensional), with time along the radial direction. The dust grains, although locally at rest on the surface, drift apart (or get closer to one another) as the expansion (or contraction) proceeds. The steady state solution, where the radius of the balloon does not change, is not allowed by the equations.

Although this result was recognized to be mathematically correct by Einstein, he did not like it because in an expanding universe, going back in time, the matter/energy density becomes infinite at a finite point in the past. When he first proposed his theory, he considered such an infinite density to be unphysical.

In order to achieve a steady-state universe, and avoid this expansion or contraction, Einstein introduced an additional arbitrary term into his equations, known as the *cosmological constant*, Λ. For a stationary solution of the equations to exist, this new constant had to be *fine-tuned*, i.e. it had to have a very precise value. The new term is perfectly compatible with General Relativity and mathematically consistent. However, any arbitrarily small perturbation of the density of matter in the universe, or of the value of the cosmological constant, produces an irreversible switch to an expanding or contracting universe. It is a little like trying to balance a pencil on its point: although theoretically possible to do, it is in practice impossible. Einstein, according to a later testimony by George Gamow, called the cosmological constant the biggest blunder of his life [2].

In a letter to Georges LeMaître in 1947, Einstein wrote [3]: "*I found it distasteful having to accept that the equation for the gravitational field must be put together out of two logically independent terms which are connected by addition. It is difficult to give arguments justifying such feelings, as I feel them in terms of logical simplicity. But I cannot fight against feeling them with all my strength, and I am not in a position to believe that such a repugnant thing could be realised in nature.*"

It may seem strange to a non-scientist to encounter concepts such as *ugliness*, and its antithesis, *beauty*, when dealing with works of science. However, as we have seen in Chap. 3, scientists are often attracted by beauty or repelled by ugliness in their theories and equations. We will examine further far-reaching implications of this tendency in Chap. 12. Einstein's dissatisfaction with his introduction of the cosmological constant may well have resided with its arbitrariness. In selecting a value to achieve a desired result, he is "tuning" his equations. The tuning of models by adjusting arbitrary parameters is a common enough practice in some areas of science, but is not generally regarded as good scientific practice.

Einstein, may have—or may not have, as there is some controversy about what he actually said to Gamow—described the cosmological constant as his greatest blunder, but that hasn't prevented his successors reintroducing it to account for *dark energy*, which we will come to in the next Chapter. Not everybody is repelled by ugliness. Einstein was very modest about his achievements and freely admitted that he had committed other blunders. In a letter to his friend and colleague, Max Born, Einstein wrote: "*I too committed a monumental blunder some time ago (my experiment on the emission of light with positive rays), but one must not take it too seriously. Death alone can save one from making blunders.*"

The last sentence is a free translation from the original German, which reads: "*Gegen das Böcke-Schießen hilft nur der Tod.*" The literal translation of this is: "Only death helps against the shooting of goats." If Einstein can so freely admit that he has shot a few goats in his life, then the rest of us should take heart. We have all of us in our various careers no doubt shot a goat or two, but most of us have tried to conceal it. Shooting goats (or the commission of blunders) is a by-product of innovative thought, even with the greatest minds in the world.

10.5 The Expanding Universe—Observational Evidence

Huge advances in the design of telescopes and in the observational techniques during the first decades of the twentieth century brought tremendous progress in the study of the sky. First, the number of visible objects formerly known as *nebulae* steadily increased. A nebula is a diffuse, cloud-like object. A well-known example, which can be observed with a small telescope or powerful binoculars, lies in the constellation of Orion, where the second "star" in the handle of the sword is diffuse. Improved resolution of telescopes revealed that these nebulae are actually comprised of many stars.

Next, an ongoing ancient debate on the whereabouts of these nebulae came to an end with the realization that they are located much farther away from us than are our surrounding stars. In fact, we now know that stars are grouped into huge conglomerations, or *galaxies*, as we call them today, from the Greek name for the Milky Way. The Milky Way, visible in our sky, is simply the inner part of our galaxy, seen from within.

Galaxies are enormous aggregates of gravitationally bound stars. They are classified according to their shape, but in general they resemble flat disks, with or without a central bulge and/or spiral arms. Astronomers believe that our galaxy, seen from outside, would look more or less like the galaxy NGC 6744, shown in Fig. 10.5.

The Milky Way is a *barred spiral galaxy*. Its diameter is in the order of 100,000 light years and our sun is located in the inner side of one of the arms of the spiral at some 27,000 light years from the centre. An observer on earth (call her Susan) looking away from the plane of the galaxy will see few stars (See Fig. 10.6). An observer (Bob) looking towards the centre will see many stars, as he is peering right through three-quarters of the full length

Fig. 10.5 The NGC 6744 galaxy photographed by the Wide Field Imager on the MPG/ESO 2.2-m telescope at La Silla. Image courtesy of European Southern Observatory (https://www.eso.org/public/images/eso1118a/ (accessed 2020/06/15))

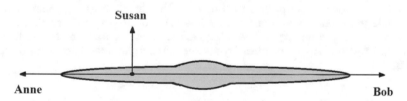

Fig. 10.6 Three observers on earth will see different densities of stars, depending on which direction they are looking out from the Milky Way Galaxy, because of the disc-like shape of the galaxy and the location of the earth in an off-centre position

of the galaxy. A third observer (Anne), looking away from the centre towards the rim, will see many more stars than Susan, but fewer than Bob, as she is looking out through only one quarter of the length of the galaxy. As we remarked earlier, the centre of the galaxy lies in the earth's southern skies.

Stars in a typical galaxy, such as the Milky Way, are counted in hundreds of billions, and within our range of sight in the universe, there are approximately a hundred billion galaxies. Figure 10.7 offers a picture taken during the ultra-deep research program of the Hubble space telescope, showing

Fig. 10.7 Ultra-deep field picture of the sky in the near infrared domain, taken by the Hubble space telescope. Apart from a few stars (displaying a diffraction pattern) of our own galaxy, what we see are all galaxies. Image courtesy of NASA, ESA and R. Thompson (Univ. Arizona) (https://spacetelescope.org/images/heic0406b/ (accessed 2020/06/15))

distant galaxies. These galaxies are photographed from their infra-red emissions, which are at a lower frequency than visible light. The portion of sky visible in the picture is approximately one tenth of the diameter of the full moon. Only a handful of the visible objects are stars of our own galaxy; the rest, even the faintest little spots, are far-away galaxies.

As we have already mentioned, the existence of galaxies does not introduce much change to the uniform, homogeneous model of the universe known as the FLRW model. Simply, instead of considering a dust of stars, the grains of the "universal dust" become whole galaxies.

To establish how far away these galaxies are from us, we need a method of measuring astronomical distances beyond the parallax method we discussed earlier, which is only applicable to nearby stars. This is an interesting problem in itself, and we discuss it in more detail in Appendix 10.1. Suffice it to say

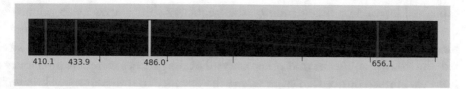

410.1   433.9   486.0   656.1

Fig. 10.8 Constructed image of bright-line (emission) spectrum of hydrogen. Other lines exist outside the visible range. The units of wavelength are in nanometres (nm); 1 nm = 1 billionth of a metre. Image courtesy of Patrick Edwin Moran (2009) (https://commons.wikimedia.org/wiki/File:Bright-line_Spectrum-Hydrog en.svg under the Creative Commons Attribution-Share Alike 3.0 Unported license (accessed 2020/06/15))

here that certain stars have a known intrinsic brightness. By measuring the observed brightness of the star and comparing it with this intrinsic brightness, we can estimate how far away the star is from the earth.

However even this approach will not be accurate for the most distant objects, and a third approach is necessary, which utilises the light spectra emitted from the distant stars. Chemical elements, when heated to incandescence, emit light of precise frequencies (or colours) specific to that particular element (see Fig. 10.8). This is known as the spectral signature of that element. We have already encountered this phenomenon in Chap. 8, when discussing the yellow emissions from heated sodium atoms.

If we were to observe the sky in the scenario of a static universe we would expect to recognize the *signature* of hydrogen (and of other elements) in the spectra of the stars. However, the stars are not stationary, and a phenomenon known as the Doppler Effect comes into play. Standing on a railway station, we will have observed a drop in the pitch of sound from an approaching train as it passes us, and races off into the distance. The wave-fronts of the sound wave emitted by the approaching train are closer together when they reach us than the wave-fronts from when the train is receding. This effect, named after Austrian physicist, Christian Doppler, is also observed for light waves, as anybody issued with a speeding ticket by the operator of a laser speed trap can testify. Studying the spectrum of the received light from the stars, it should be possible to recognise the signature of hydrogen, but we would expect the light to appear redder or bluer than in a corresponding terrestrial laboratory experiment due to the motion of the stars. In a static universe, we would expect to find approximately equal numbers of stars approaching us as receding from us.

In the case of an expanding or contracting universe, we expect to observe a similar colour shift in the spectra of stars, analogous to the Doppler Effect, which we have just described. (It is not exactly the same effect, as will be explained later.) The magnitude of the frequency shift allows us to calculate the velocity of the source away (or towards) us. Carrying out these measurements with galaxies, Hubble found that in almost all cases, the light was red-shifted, which meant that practically all of the galaxies were receding away from us. He also noticed, in cases where the distance to the galaxy was known, that the red shift was proportional to the distance. The constant of proportionality is now known as Hubble's constant. (Note the implicit assumption here that the expansion is uniform.) From Hubble's constant and the observed red shift we can calculate the distance to the farthest astronomical objects. So how do we estimate Hubble's constant?

Fortunately, there is a region of overlap between where luminosity measurements can be used to estimate distance, and where red shifts of the closest galaxies can be measured. A comparison of the two sets of observations enables us to obtain an estimate of Hubble's constant. It sounds simple, but decades of work have been undertaken to refine the accepted value of this constant. These estimates have fluctuated quite considerably. The estimated age of the universe is directly related to the Hubble constant.[2]

Measurement of the red-shift of galaxies became a turning point in the history of Cosmology. Einstein abandoned his cosmological constant, and the FLRW solution of the field equations of General Relativity lost its status as a mathematical curiosity and became accepted fully-fledged into the domain of physics. The expansion of cosmic space became generally accepted.

Earlier we considered a sphere as an analogy to the four dimensional universe. We develop this analogy further by considering the sphere to be an inflatable balloon, with the stars (and galaxies) as specks on the balloon's surface. As the balloon expands, the specks move further apart from each other. Note that the specks do not move across the balloon's surface, but rather the surface expands between the specks. Similarly, in the four-dimensional universe, the galaxies do not move through space, but space expands between them, making them further apart. This is what we meant earlier when we said that the red shift of the expanding universe was *analogous* to a Doppler shift, but not exactly the same. An important difference is that whereas the Theory of Relativity (see Chap. 9) restricts the velocity of stars moving through space to be less than the velocity of light, the stellar motion resulting from the cosmic expansion of space itself faces no such limitation. As we shall see in a later Section, this difference has important implications in the early life of the universe.

---

[2]Hubble's law had been anticipated a couple of years earlier by Georges Lemaître, while analysing the implications of his expanding universe.

An immediate consequence of the expansion of space was the *finite age* of the universe, the current estimate of which is 13.8 billion years. If we could go backwards in time in such an expanding universe, we would find the mass/energy of all galaxies becoming increasingly squeezed by the reduced space that was available to it at these earlier times. The density of matter/energy would steadily increase until at the origin of the universe we would presumably encounter an infinite density condition, which we might call a *space–time singularity*.

Before moving on, it is opportune to discuss in more detail what we mean by *singularity*. Singularities in mathematics arise when some variable takes an infinite value. For instance, consider the reciprocal $1/x$ of a number x. The smaller the number, the larger the reciprocal. The reciprocal of 0.1 is 10, of 0.01 is 100, of 0.0001 is 10,000, and so on. Try taking the reciprocal of 0 with your calculator and you will obtain an error message; this is the calculator's way of saying the answer is infinity and handling infinities is outside its job description, thank you very much.

Singularities are common enough in mathematics, but physicists don't like them. Remember Einstein introduced his cosmological constant to avoid the very singularity that we are discussing above. Sometimes singularities bound a region that we cannot enter. For instance, we have seen in Chap. 6 that when we increase our speed, the closer we get to the velocity of light, the greater our mass becomes. If we could ever travel at the velocity of light our mass would become infinite, which indicates that it is impossible for us ever to attain that speed. In other cases, as we get closer to a singularity, the laws of physics that we are applying break down, and we need to find new laws that are applicable in this particular domain.

Upon hearing that the universe has an origin, the first *naïve* question that comes to mind is "what happened *before* the origin?" Such a question does not make sense, and is contradictory. The word "before" implies time, but the singularity is the origin of time, as well as of space. There is no "before". Of course, physicists did not (and still do not) feel at ease with the idea of an origin of time, since it invokes scenarios outside the domain of science. This point is discussed further in Part 3.

Although the red shift has now become a routine tool in Astronomy, not everybody accepted its interpretation in terms of the expansion of the universe, and several scientists proposed alternative explanations, since abandoned. In 1948 British scientist, Fred Hoyle, proposed a mechanism that would have guaranteed, in his opinion, a *steady state* for the universe. (Independently, similar ideas were also considered by Hermann Bondi and Thomas Gold.) Hoyle accepted that the cosmic red-shift was due to an expansion, but rejected the corollary of a hotter past and cooler future. In particular, he rejected the idea of an initial singularity, which he jeeringly nicknamed "*big bang*", never suspecting that the term would soon become a hit, even among the supporters of the FLRW model. Hoyle maintained that a steady state universe could be compatible with the drifting apart of galaxies if, due to a "*creation field*", *new* matter were spontaneously and continuously generated in the intergalactic space. The amount of newly created matter required to ensure the steady state condition was very little: approximately one hydrogen atom per cubic kilometre per year. This is so small as to be totally unobservable.

The steady state theory was thus based on the assumption of an unobservable phenomenon, and as such might be thought incompatible with what we call "physics". (Theories in physics should produce some predictions that are capable of being verified or refuted experimentally, either directly, indirectly, now or in the future.) Wolfgang Pauli, one of the greats of 20th Century Modern Physics, disparaged untestable theories as "not even wrong". This, in his eyes, was a far worse characteristic than being wrong, for the experimental testing of wrong theories often leads to unexpected new breakthroughs. However the Steady State Theory does predict observable differences with the FLRW model. Newly discovered radio sources (quasars and radio galaxies) were associated in the big bang theory with the early stages of the universe; they were therefore expected to be found only at large distances.

From the Steady State Theory, these unusual new objects were expected to be uniformly distributed throughout the universe. This disagreement between the Steady State Theory and the FLRW model led Steven Weinberg to write in 1972: "*In a sense, this disagreement is a credit to the model; alone among all cosmologies, the steady-state model makes such definite predictions that it can be disproved even with the limited observational evidence at our disposal* [4]."

The steady state theory was finally swept away by new observational evidence emerging in 1964 and analysed and perfected afterwards: namely, the *relic radiation* or *cosmic microwave background* (CMB), which we discuss in the next Section. Hoyle's reluctance to accept the demise of his theory shows that even a front rank astrophysicist like Fred Hoyle may not be immune from prejudices.[3]

---

[3]The Steady State Theory received a new incarnation in the Quasi-steady state cosmology (QSS) proposed in 1993 by Fred Hoyle, Geoffrey Burbridge, and Jayant V. Narlikar. It was intended to explain additional features unaccounted for in the initial proposal, but ran into further difficulties and is not generally accepted.

10.6 Cosmic Microwave Background

Now that we have established that our universe began with a Big Bang about 13.8 billion years ago, and has been expanding ever since, it is time to look at some of the events that have occurred along that timeline. We will discuss the earliest moments of the universe in the next Section, but first let us begin our story with an important period, the so-called *recombination era,* about 380,000 years after the Big Bang, and in so doing resolve Olber's Paradox. At that time, according to the FLRW model, matter was in the form of a rarefied plasma and the estimated temperature was around 3000 K. A plasma in a terrestrial experiment is created when matter is heated until the temperature is so high that atoms cannot exist, but are split by the thermal energy into their constituents. Before the recombination era, charged particles (mostly electrons and protons) could not bind into atoms because the thermal radiation they continuously emitted and absorbed had enough energy to prevent the formation of stable bonds. The cosmic medium at that time resembled a bright white fog.

Afterwards, as the temperature decreased below the threshold of 3000 K, the average energy per photon became insufficient to break the atomic bonds when they formed, so that hydrogen atoms started to appear. Also photons, not having enough energy to ionise hydrogen, (i.e. strip its electron away from the atom) could no longer be absorbed, but only scatter elastically and the universe started to become transparent.

The light emitted from the hot plasma of the recombination era had a distribution of frequencies, or a spectrum, that is characteristic of what is called "black body radiation". It may seem somewhat odd to call a white-hot, incandescent mass a "black body". The terminology arises because when cold, the object is a perfect absorber of incident radiation, i.e. it is black. This is an ideal situation, because most objects reflect some of the incident radiation, and are therefore not perfect absorbers. As we have seen in Chap. 5, the attempt to explain black body radiation resulted in the birth of Quantum Mechanics.

We expect light escaping from the recombination era and arriving at earth today to display the type of black body radiation spectrum discussed in Chap. 5. However, if the light is not absorbed, due to the expansion of space it will undergo a red shift similar to the red-shift of light from stars in distant galaxies, which we discussed in the previous Section. In this case, however, the distance is so great that the radiation wavelength is stretched by a factor of approximately 1100. What was initially visible light now appears as microwave radiation, and is known as the Cosmic Microwave Background

(CMB). It displays a typical thermal radiation spectrum, in which the initial 3000 K radiation spectrum of the primordial hot plasma is red-shifted to one corresponding to a temperature T_{CMB} of just 2.72548 K.

In 1964 Arno Penzias and Robert Wilson, while testing a microwave antenna at the Bell Telephone Laboratories, accidentally found a "noise" coming uniformly from every direction in the sky. That "noise" was quickly recognized as being the cosmic microwave background, a.k.a. the *relic radiation*, arriving to us after having survived more than thirteen billion years of travel through space–time. Penzias and Wilson received the Nobel Prize for their discovery in 1978. Their experimental observation sounded the death knell for Hoyle's steady state universe.

Also, now at last Olber's paradox is explained. The sky *is* bright at night. However, it appears black because our eyes sense only a very small portion of the electromagnetic spectrum, and the white light from the recombination era has been red-shifted into the microwave region where our eyes see nothing. For a microwave antenna, the sky is uniformly bright, not as the surface of a suitably red-shifted star, but as a red-shifted hot plasma transitioning to a neutral gas.

The uniformity of the CMB, both in intensity and temperature is amazing, with fluctuations of only 1 part in 100,000. This uniformity raises an interesting question. As we have seen, the first 380,000 years of the universe's life are impenetrable to our observations. When we look back in time through our telescopes,[4] our view is stopped, as if by a curtain, by the recombination taking place at this time, at what is known as the *last scattering surface*. However, distant portions of the sky, from which the radiation comes, were at that time separated by more than 380,000 light years. As no physical "messenger" can travel faster than light, no causal interconnection can have taken place between these far-flung regions since the Big Bang. Then how is it possible that these independent regions can have kept such an amazing synchrony, arriving at the same average temperature in the same time? We shall come back to this point in the next section.

Yet differences among various areas of the sky do exist. Evidence of this inhomogeneity can be seen in Fig. 10.9, where the temperature of the radiation coming from the whole cup of the sky is shown. To be sure, the unprocessed data from the sky are not as clean as in Fig. 10.9, since there are also microwave emitters in the foreground, primarily from the Milky Way. In order to study the primordial radiation field, these nearby sources had to be

---

[4]The light our telescopes receive from distant objects has taken a long time to reach us. What we see is the situation as it was when the light was emitted, and not as it is now. Our telescopes are therefore "looking back in time."

Fig. 10.9 Distribution of the equivalent temperature of the CMB across the sky. False colours have been used: red means warmer; blue means cooler. The measurement has been made by the NASA space-based microwave telescope WMAP. Image courtesy of NASA/WMAP Science Team (https://map.gsfc.nasa.gov/media/121238/ilc_9yr_moll4096.png (accessed 2020/06/15))

removed from the data. The picture, taken from the NASA WMAP survey, presents the temperature of the radiation using false colours: the red areas are the warmest, the dark blue the coolest. Remember that the difference between the coolest and the warmest is less than 1 part in 100,000, so the differences have been enormously magnified in order to make them perceptible to the eye. To understand the figure, imagine being at the centre of the celestial sphere, cut it from one pole to the other, then open and flatten it. In this way, the right border corresponds to the left one.

Despite their smallness, the anisotropies in the CMB carry a wealth of important information about the primordial universe. An immediate remark, looking at Fig. 10.9, is that the distribution of warm and cool areas is not really random: some irregular structures do seem to exist. First, there is a concentration of red and yellow blots on the right side of the image and an area of prevailing dark blue in the middle (and again on the right border). Returning to the initial three-dimensional sphere, the warmest and coolest areas turn out to be in opposite directions in the sky: this is the *cosmic anisotropy* of the CMB.

The explanation is simple. If the earth were stationary with regard to the space surrounding it, we might indeed expect the observed red shift to be the same in all directions. In our balloon analogy, in this case the speck representing the earth does not move across the surface of the balloon; the balloon expands carrying the earth with it. However, the real earth is not stationary, but moves around the sun, the sun moves within the galaxy, and the galaxy moves through the cosmos. This motion introduces a further shift of frequencies (red or blue, depending on the direction of the earth's motion relative to the incoming radiation) superimposed on that of the CMB. This additional shift is a true Doppler frequency shift due to the earth's motion through space, unlike the cosmic red shift due to the expanding universe. From the observed anisotropy in the CMB relic radiation, the "absolute" velocity of the earth (i.e. the velocity with respect to the background) can be estimated, and is found to be 371 km/s in the direction of the constellation Leo.

Once we know its origin, we may as well subtract this Doppler anisotropy, so that we are left only with the primordial anisotropies. The study of the latter is extremely important in many aspects, which we cannot discuss here. However, a quick qualitative remark is that the tiny density fluctuations present at the recombination era are the "seeds" of the future galaxies. In fact, any density excess, via a gravitational positive feedback, would tend to grow, attracting matter from the surrounding less dense regions and in the process give rise to large scale structure.

Also, the pattern of density fluctuations of the last scattering surface carries the imprint of what happened *before* that era. Nothing else can arrive to us from earlier times closer to the Big Bang. The last scattering surface acts as a bright impenetrable curtain that prevents us from plunging deeper towards the singularity. The history of the universe before the recombination may be inferred to some extent from the accepted laws of physics, but not observed. This is certainly true for the information carried by known physical interactions. Various non-standard theories predict the existence of other "messengers" from the very early times of the universe. Besides these, it has been conjectured that gravitational waves could also exist as a relic of the pre-recombination era. However at present no observational evidence exists for any of this.

10.7 The Universe, as It Was Known at the End of the 20th Century

Let us now assemble facts and theories to construct a consistent cosmology, as it was accepted at the end of the twentieth century. General relativity was the basic conceptual framework, and the expanding universe had been well established in the FLRW formalism. A consistent chain of physical events may be described using known physics (including quantum mechanics for nuclear and sub-nuclear forces) from very close to the initial Big Bang up to the present time. Not everything is clear and understood, but the fundamental pillars are well established.

The initial singularity of the Big Bang remains outside the domain of physics, since nobody knows what to do with an infinite energy density. Immediately following the Big Bang, we don't even know what the laws of physics were. At these incredible energies, the four forces may well be unified into a single force. We expect gravity to play a dominant role because of the enormous densities, and consequent curvature of space–time, in this region, but so far it has stubbornly resisted all of our attempts to bring it into the framework of quantum physics.

Despite these difficulties, theorists have proposed a mechanism to explain the puzzling homogeneity of the Cosmic Microwave Background that we discussed in the previous Section. Allan Guth and Andrei Linde in the 1980s conjectured that in the period leading up to 10^{-32} s after the Big Bang, the universe underwent an exponential expansion. This might seem an inconceivably short time period, but the Big Bang was proposed as initiating at a singularity, which is, as we have seen, a *point* with *zero* volume in space–time.

Such was the expansion rate that the scale of spatial distances increased by a factor of at least 10^{26}. To achieve such an enormous rate, the recession velocities of the components of the universe, would have been well above the speed of light. As discussed earlier, these superluminal velocities are not a contradiction of relativity, because it was space–time itself that was expanding. This period of expansion was called *inflation* by Guth and Linde.

The development of inflation theory is highly mathematical, and beyond the scope of this book. However, a few qualitative observations are possible. The period of exponential expansion smooths out any density fluctuations present at the beginning of inflation. As a consequence, the CMB is incredibly homogeneous, as we have already seen. However, tiny residual inhomogeneities (see Fig. 10.9) do remain. Once the inflationary expansion is over, these inhomogeneities are frozen into the fabric of the universe because the regions containing them can no longer interact with each other. Such interactions would require faster-than-light travel, which is forbidden by Relativity.

To explain this period of inflation, Guth and Linde proposed a new field, called the *inflaton,* to describe matter/energy. As the expansion proceeds, the energy density decreases until the inflaton field decomposes, and the more familiar fields (strong nuclear interaction, weak interaction, and electromagnetism) decouple from each other. The end result is a type of phase transition resulting in the creation of the particles (gluons, quarks, etc.) carrying the above interactions. A phase transition is a common phenomenon in physics, where an entity can exist in different forms (e.g. as H_2O can exist as ice, water and water vapour), and under particular conditions can transform from one to the other.

After the phase transition, the cosmic expansion continues at a much slower rate, and we enter a domain where we expect more familiar physics to prevail. The quarks and gluons begin to form protons and neutrons (hadrons). As the temperature falls, these hadrons combine to form the nuclei of atoms, such as deuterium (one proton and one neutron) and helium (two protons and two neutrons). The nucleus of deuterium is known as the deuteron, and the nucleus of helium is the α-particle, familiar from the earliest studies of radioactivity. The proton itself is the nucleus of the hydrogen atom.

At this stage an opportunity arises to use existing nuclear theory to predict the relative abundances of these lightest elements, as well as the slightly heavier ones of lithium and beryllium, in the primordial universe. Although the issue is not yet completely settled, the calculated relative abundance of helium (25%) is in good agreement with observation.

The next significant stage in the evolution of the universe is reached when, after approximately 380,000 years, the temperature drops sufficiently to allow electrons and atomic nuclei to combine to form neutral atoms. Before this time, the free electrons interact with any photons to prevent their passage. Once the electrons have been attached to nuclei to form atoms, their electromagnetic field is largely countered by the oppositely charged nuclei, and their ability to impede the passage of photons is greatly reduced. The universe becomes transparent, and visible through large telescopes on present-day earth, looking back through time and space to this past era.

With the short range nuclear forces contained within nuclei, and the electromagnetic force reduced in potency by the close proximity of oppositely charged particles, gravity becomes the driving force in the universe. The tiny inhomogeneities that survived inflation begin to grow under its influence. Slightly dense regions become denser as they attract more particles into their neighbourhood, with a resultant increase in temperature and pressure. The temperature rises in these regions until the nuclear fusion of hydrogen into helium begins, and the first generation stars are born. These are powered by the "burning" (nuclear fusion) of hydrogen.

From Fig. 10.9 we can infer that the scale of the inhomogeneities is much larger than the dimensions of stars. The stars themselves are subject to gravitational attraction, and group themselves slowly into clusters, or "galaxies". Even the galaxies may become clustered. The time scale over which significant changes now manifest themselves is measured in millions (or billions) of years.

Stars pass through a life cycle of their own, converting their hydrogen into deuterium and helium, and depending on their mass, may commence burning their helium, converting it into heavier elements, up to iron and nickel. During this phase of nuclear fusion inside a star, a balance is achieved between the outwards radiation pressure from the fusion process, and the gravitational collapse of the outermost layers of matter. Following the exhaustion of the nuclear fuel in heavier stars, the radiation pressure falls until it can no longer oppose the gravitational forces. The core collapses, and there is a resultant huge release of energy. During this period of core collapse, nuclei heavier than iron are synthesised. A supernova, a star of immense brightness, appears in the sky for several days, or weeks. The outer layers of the star, containing traces of the heavier elements, are blown away into space, and the core collapses further into either a neutron star or a black hole. (Neutron stars and black holes have been discussed in Chap. 7.)

From the beginnings that we have outlined above, the universe has evolved into the amazing variety of objects, including planets and life-forms, that abound today. It is humbling to realise that the atoms, other than the very lightest, from which our bodies are constructed, were forged in the furnace of a star. This is the domain of astrophysics and astronomy.

What we have presented above is the standard picture of the universe as it was two decades ago. As often happens in physics, when physicists believe they are consolidating their theories and obtaining a glimmer of understanding, new discoveries occur that bring everything into question once more. In the next Chapter we will discuss some of these new observations and ideas, and the grey areas and open puzzles that require further investigation.

References

1. West M (2003) The last confession. Toby Press
2. Weinstein G, George Gamow and Albert Einstein: Did Einstein say the cosmological constant was the "biggest blunder" he ever made in his life? (October 3, 2013) https://arxiv.org/ftp/arxiv/papers/1310/1310.1033.pdf. Accessed 14 June 2020

3. Einstein to Lemaitre, 26 September 1947, Einstein Archives Doc. 15 085. Kragh, Helge, Cosmology and Controversy: *The Historical Development of Two Theories of the Universe,* Princeton: Princeton University Press, 1996, p 54

4. Weinberg S (1972) Gravitation and cosmology. Whitney

11

Modern Cosmology

11.1 Hyperspace and Wormholes

Binh had been very apprehensive when he emigrated from Vietnam to Sydney with his young family. It was okay for him. He was a research fellow in astrophysics at a prestigious university, but his son had to adjust to a new school, and learn English before he could even begin his courses. However, Binh needn't have worried. Danh was a bright boy and had soon made lots of friends. He'd also begun to pick up some of the local bad habits, including an over-fondness for video games. Binh was pleased when Danh's English became good enough for him to read books. He'd bought him a few on the history of his old country, but Danh preferred adventure novels. He burst into Binh's study carrying one.

At the unexpected entry, his father looked up from his desk. "What are you reading now?" he asked. Danh handed him the book. It was *The Little Star Gatherer*. "Any good?".

"Good? It's wicked."

"Wicked, eh?" Binh smiled at the strange expression.

"There's this boy called Rajveer. He's got a starship so fast he can go to other solar systems and visit their planets. And every planet is different, with all sorts of animals and plants. And aliens."

"Who all speak English, no doubt?".

R. Barrett and P. P. Delsanto, *Don't Be Afraid of Physics*, https://doi.org/10.1007/978-3-030-63409-4_11

"Is it really possible to visit other planets, *ba*?".

"You mean planets in other solar systems?" Danh nodded. "They're called *exoplanets*. Well, let's see if we can work it out, shall we?" Danh settled into the spare chair. He'd been through these didactic sessions with his father before, but he loved him too much to tell him how boring they were. "The nearest star system to us is *Alpha Centauri*, and yes, you're in luck. It does have an exoplanet. But it's four-and-a-half light years away. So even if you could travel at ten percent of the speed of light – which is unthinkable – it would still take you 45 years to get there."

Danh pointed at the novel in Binh's hand. "In there, Rajveer travels through hyperspace."

"Does he now?".

"What's hyperspace, *ba*? Is it real?".

"Now that's a very good question." Danh waited for an answer. It was the problem that he'd been pondering over all morning, and was the reason he'd come into his father's study. Binh picked up a sheet of paper. Danh's heart sank further. "Let's imagine we are two-dimensional animals living in a two dimensional universe, like this sheet of paper."

"Does it have anything to do with wormholes?" The boy's interjection stopped Binh short, just as he was about to launch into a full explanation, and demonstrate with a folded paper how a two dimensional universe could be curved in a third dimension.

"What do you know about wormholes?" Binh was most impressed. He thumbed through the pages of The Little Star Gatherer to see what other pearls it might contain.

"A bit. But are they real?".

"Maybe. Maybe not."

"*Ba!* That's no answer!".

"No, but it's the only one I can give you." In truth Binh, who prided himself on his teaching skills, was a little annoyed that his son had interrupted his discourse before he had even got going. "If you want to know more, then you'll just have to study physics when you get to high school, and then go on to university. And stop wasting so much time on video games."

"Same old, same old," muttered Danh under his breath. He did like video games, but he also liked science, and his grades were good. But his father was away again on his favourite hobby horse.

This time, however, the boy had a prepared response. He reached across for *The Little Star Gatherer*, and rose from his chair. At the door, he turned back towards his father, and called: "Hey, *Ba*, did I tell you our sports teacher thinks I should try out for the football team? He said I show a lot of promise. Never know. Might even get good enough to be a pro one day." He closed the door quickly to hide his grin, and strode smirking down the passage to his own room.

* * *

Danh's question is a good one. Wormholes, or tunnels connecting different events in space–time, are valid solutions of Einstein's field equations. They do however raise the problem of time travel, where travellers might go back into the past and kill their parents, thereby threatening their own existence. Perhaps fortunately, there is no experimental evidence for the existence of wormholes.

11.2 The First 380,000 Years

In the last Chapter we have been diligent in our attempt not to stray too far from established theories in a search for an explanation of what is arguably the oldest and most puzzling question ever to have confronted humanity: *how did the universe in which we live begin?* This question is basic for all religions: physics cannot simply ignore it, and take refuge behind the defence that a direct observation of the Big Bang is impossible. However, when venturing to explore this region, caution is necessary, controversy unavoidable, and frankness laudable.

Although many people are working diligently in this field, it is openly admitted that the physics immediately following the Big Bang is completely unknown. Unfortunately, that period contains the original central singularity at the origin of the universe. It is where the two main cathedrals of modern physics, Quantum Mechanics and General Relativity—one probabilistic and the other deterministic in nature—are mutually inconsistent. We have discussed this conflict already in Chap. 9. Physicists have tended to skate around this issue, applying Quantum Mechanics when dealing with small particles, and General Relativity when gravity is the dominating force.

However, at the Big Bang, and shortly thereafter, the disagreement must be confronted head on. There is no place to hide.

As an aside, let us remark that similar issues arise wherever singularities appear. Black holes are examples of singularities predicted by physical theory, and the physics close to the central singularity in a black hole is not understood for the same reason that we discussed in the preceding paragraph. In this region of intense gravitational attraction, the theories of Quantum Mechanics and General Relativity are in conflict, and the black hole is surrounded by an event horizon, from within which no light or other particles can emerge. Below the horizon, a *cosmic censorship* principle is acting: we cannot see in there, so it is regarded by many as beyond the purview of physics. Yet physicists have accepted the existence of black holes. For example, observations tell us that a massive object, compatible with a black hole, has been found at the centre of our galaxy. It has been named Sagittarius A*, and has a mass equal to four million solar masses. It is now believed that black holes are located at the centres of most galaxies.

In the last Chapter, we discussed the *inflation* model, and saw that it was necessary to introduce an entirely new field, the *inflaton* field, to explain the extraordinary homogeneity of the Cosmic Microwave Background. This uniformity could not be explained by conventional theories because, even over 13 billion years ago, the most distant parts of the universe were too far apart to be accessed, one from the other, at velocities less than the velocity of light.

Here, we are assuming that the velocity of light at this time was the same as it is today. This is in accord with our belief that the physical constants do not change, which we discussed in Chap. 3. Suppose, however, that our belief is wrong and that the speed of light in these early times was much larger than it is today. A Variable Speed of Light (VSL) model of early cosmology, as an alternative to the inflation model, was proposed by John Moffat in 1992 [1]. This work was largely ignored, until in 1998 another physicist, João Magueijo, published a similar idea in a more prestigious journal. An unpleasant controversy erupted over which author had priority, with the media largely ignoring Moffat's work. Eventually the two physicists were reconciled, and have since published further papers jointly. Light speeds up to 60 times the current value of c have been suggested. A VSL model would have implications in special and General Relativity, Maxwell's equations of electromagnetism and many other areas of physics.

Currently VSL models of early cosmology are outside the mainstream of physics. However, the inflation theory is also not without its critics. Paul Steinhardt, one of the original contributors to inflation theory, is now concerned that the model requires fine tuning to explain the universe, as we know it. By varying parameters in the model, *any* conceivable universe can be explained. Steinhardt goes further and adds that inflation would produce not just one universe, but a multiverse containing all possible universes with all possible properties. There is no way to verify, or falsify, such a theory, which Wolfgang Pauli might well say is "not even wrong" (see Chap. 10). Steinhardt's (and Pauli's) rejection of non-testable theories is not a philosophical viewpoint shared by all physicists. Many take a different view, i.e. that a theory can simply provide an explanation that relates observations to underlying physical principles.

Steinhardt and collaborators have proposed that instead of inflation, the universe is cyclic in nature, and our universe has arisen from the remains of an earlier collapsed universe. This approach is not without its own difficulties, running afoul of the Second Law of Thermodynamics, which requires that entropy (or disorder) increase inexorably. A baby universe is in a low-entropy ordered state, and an old universe is in a much higher entropy state. A recycled universe would begin its life in an unacceptably high entropy state. Roger Penrose and others have made suggestions on how to overcome this problem by modifying some of the laws of physics.

In this section, we have probed into some of the controversy that underlies the physics of the early universe. We conclude by noting that the inflation theory, as we described it in Chap. 10, remains the most popular of the contending theories among astronomers and cosmologists. It is perhaps also worth giving ourselves the wry reminder that the most popular cosmological theory of the 1950s was the Steady State Theory, long since defunct. In the long run, popularity in science counts for very little.

11.3 The Strange Case of the Missing Mass

In the 1930s, from studies of the orbital speed of galaxies in clusters, and stars within galaxies, including the Milky Way, it was discovered that the velocities of objects gravitationally bound to each other were higher than expected [2, 3]. The observed velocities were explained by assuming that more mass existed in the galaxy, or the galaxy cluster, than was actually observed. We have noted already in Chap. 3 that interstellar dust is a feature of our galaxy. Matter is also present in the remnants of dead stars, planets and asteroids, none of which are visible from their radiation. However the sum total of all such obscure matter is estimated to contain less mass than the visible stars. In order to explain the anomalous orbital velocities, it would appear that there is a *missing mass* in the universe, and that it must be more than five times the visible mass.

Today, the term *dark matter* has been coined for this missing mass. If dark matter is assumed, the anomalous motion can be explained within the framework of General Relativity. The actual nature of the dark matter, however, remains a mystery. Many conjectures have been made about the components of dark matter (e.g. neutrinos and other exotic particles), but none have been satisfactorily substantiated.

Alternative explanations of the anomalous motion of stars and clusters involve modifications of the theory of General Relativity. One such theory is MOND, or Modified Newtonian Dynamics, created in 1983 by Israeli physicist Mordehai Milgrom [4]. In this theory, Newton's law departs from the well-known inverse square law when the accelerations involved are very small, such as in the outer reaches of galaxies. The modification to Newton's theory is too small to be observed on earth or within the solar system, where the accelerations are much larger. MOND has been successful in explaining a number of galactic phenomena, but fails to predict the behaviour of galactic clusters. It has attracted a small number of adherents, whereas the majority of cosmologists are committed to the dark matter solution of the observed anomalies.

Recent evidence of dark matter has been found by Seth Epps and Michael Hudson of the University of Waterloo [5]. Current theory predicts that galaxies are immersed in a halo of dark matter, and that galaxies that are relatively close to each other are connected by filaments of dark matter. Light passing in the vicinity of any matter is bent by the curvature induced in space–time by the presence of the mass. The effect is known as "gravitational lensing" (see Chap. 7). Epps and Hudson averaged the results of gravitational lensing from 23,000 pairs of neighbouring galaxies, and compared these with similar averages from pairs of galaxies that were in the same region of the sky, but actually well separated in distance.

Results from their study are shown in Fig. 11.1. A red bridge associated with dark matter filaments between the pairs of neighbouring galaxies (white regions) is evident in the upper false-colour image, but not in the lower image, which displays the results from well-separated galaxies.

Further investigations are needed to confirm the results of Epps and Hudson: meanwhile the search for the elusive dark matter continues.

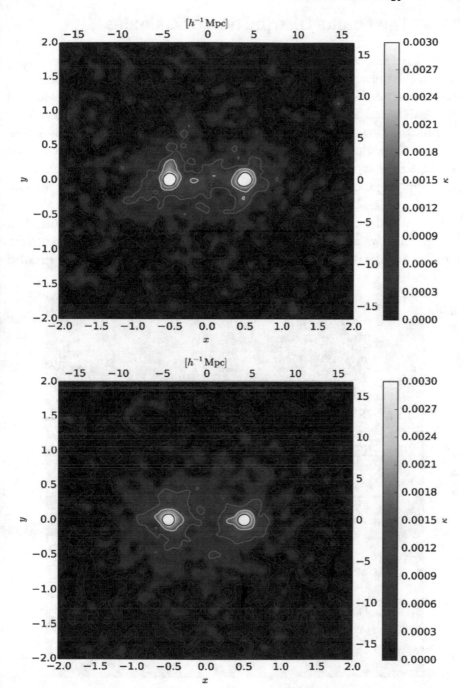

Fig. 11.1 Gravitational lensing between pairs of galaxies. The upper image shows a bridge between neighbouring galaxies which is absent in the results from well-separated galaxies (lower image). *Image courtesy* OUP (Image source: Seth D. Epps, Hudson, Michael J. *The weak-lensing masses of filaments between luminous red galaxies* (Mon. Not. R. Astron. Soc. 468 (3): 2605–2613, 2017, Fig. 3)

11.4   The Foamy Distribution of Galaxies

As we have seen earlier in our discussion of Olber's Paradox, the prevailing view of the universe in the nineteenth century was a uniform distribution of stars throughout space. Subsequently it became clear that stars are grouped in galaxies, and it was presumed that these galaxies were homogeneously distributed. However, we now know that the galaxies are distributed in clusters and these clusters in superclusters. Looking at the broadest picture we still expect the superclusters to be spread homogeneously. In this belief, we are again applying Occam's Razor, for a homogeneous distribution is one without structure, which involves fewer assumptions than one with structure.

However, the evidence from observation is not in accord with this view. In Fig. 11.2 we have a slice of the sky, centred on the earth, taken from the Sloan Digital Sky Survey. This survey has created extremely detailed

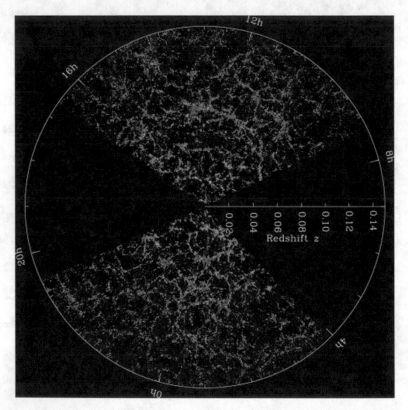

Fig. 11.2 Distribution of galaxies in a slice of sky, centred on the earth. The picture is taken from the Sloan Digital Sky Survey. Image courtesy of M. Blanton and the Sloan Digital Sky Survey (https://classic.sdss.org/includes/sideimages/sdss_pie2.jpg (accessed 2020/6/18))

three-dimensional maps of the Universe, and contains spectra for more than three million astronomical objects. It measures red-shifts using a dedicated 2.5-m wide-angle optical telescope at Apache Point Observatory in New Mexico, U.S.A.

We see from Fig. 11.2, in which every little dot is a galaxy, that the observed distribution of galaxies is anything but homogeneous. Close inspection reveals a spongy or foamy structure. The galaxies are located on walls and filaments that border large voids where nothing is visible. The underlying cause of this structure is unknown. It may be related to events that arose in the pre-recombination era, or to the prevalence of dark matter. Speculations abound, but we really don't know.

11.5 The Accelerated Expansion

In Chap. 10, we discussed the understanding of the cosmos at the stage that it had reached towards the close of the 20th Century. However, at the end of 1998 and at the beginning of 1999, the complacency that was becoming common among cosmologists was shattered when two independent groups of astrophysicists, guided respectively by Adam Riess and Saul Perlmutter, published the results of extended surveys of type Ia supernovas located in other galaxies.

The measurement of distance in astronomy by the "standard candle" approach is described in Appendix 10.1. Type Ia supernovas are suitable standard candles; they all have the same intrinsic brightness, which can be estimated by measuring their apparent brightness at known distances. Further measurements of the apparent brightness of these supernovas in more distant galaxies can then provide an estimate of the distance to these galaxies.

A second independent measurement of this distance can be obtained by measuring the red shift from one of the spectral lines visible in the light emitted by a supernova, and applying the Hubble constant to calculate the distance to the supernova, and thus to the galaxy that contains it. Unfortunately, as is sometimes the case in physics when two independent approaches are available to measure the same quantity, the results from the two sets of measurement do not agree with each other. It is found that for type Ia supernovas at distances larger than approximately 1 billion light years, the supernovas appear to be fainter than they should be. Since both types of data come from the same sources, the explanation cannot be in terms of different distances.

A solution to this puzzle has already been hinted at in Sect. 10.5: when calculating the distance to a supernova by measuring the red shift and using the Hubble constant, the assumption is made implicitly that the Hubble constant is a *constant*. Its very name is an indication of our confidence in this assumption, which is an application of Occam's Razor. The data from the two sets of measurements suggest that the expansion rate of the universe at the time of emission of the light (measured by the red shift) was less than it is today. The higher expansion rate in the universe today means that the path length travelled by the light to reach us (on which the apparent luminosity of the supernova observed through our telescopes depends) is longer than one would expect from the initial expansion rate. In other words: *the expansion of the universe appears to be accelerating*.

So how do these new results accord with Einstein's Theory of General Relativity, which is the theory underlying most of our understanding of cosmology? Not very well at all, as it turns out. In fact, depending on the actual value of the average energy/matter density in the universe, General Relativity admits three different solutions, but in all cases the gravitational attraction between galaxies *slows down* the initial expansion.

In the first, high density solution, the expansion reaches a maximum and the universe begins to contract back, ending in a final Big Crunch. The Big Crunch can be considered to be the inverse of the initial Big Bang singularity. This solution constitutes a closed universe. (See Fig. 11.3.)

The second solution occurs when the energy/matter density has a value of approximately five hydrogen masses per cubic meter. This value is critical, and even a small deviation from it will not lead to this solution of Einstein's equations. In this solution, the expansion slows down and stops asymptotically at an infinite cosmic time. This would result in a flat and open universe.

In the third solution, which occurs for smaller densities, the expansion rate also slows down, but tends to an asymptotic value different from zero, so that the expansion never ceases. This would result in an open, but not flat universe. As we can see, under no circumstances do the equations of General Relativity yield a solution where there is an acceleration of the expansion rate. The conclusion is clear: either General Relativity is wrong, or we are missing something in our understanding of the universe.

In the preceding paragraphs we have used the terms "open" and "closed" to describe the universe, and "flat" and "curved" to describe space–time. What we mean by these can be illustrated by Fig. 11.3. The upper figure corresponds to a closed universe with curved space–time. The parameter Ω_0 describes the energy/mass density in arbitrary units. In this space, Euclidean geometry does not apply, and the angles in a triangle sum to more than 180°.

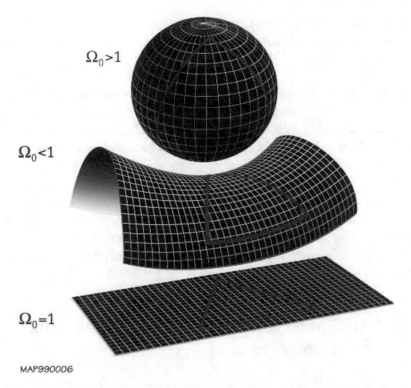

$\Omega_0 > 1$

$\Omega_0 < 1$

$\Omega_0 = 1$

MAP990006

Fig. 11.3 Three different representations of the universe corresponding to a closed, positively curved universe; an open, negatively curved universe; and an open, flat universe. Image public domain, Courtesy: NASA/WMAP Science Team (https://com mons.wikimedia.org/w/index.php?curid=647033 (accessed 2020/6/19))

The space is said to have positive curvature, and corresponds to the first solution in the preceding paragraph. The middle figure corresponds to an open universe with negative curvature of space–time. Here the geometry is non-Euclidean and the angles of a triangle sum to less than 180°. This corresponds to the third solution of the General Relativity equations. The lower figure is an open universe with flat space–time. It is the second solution from the preceding paragraph. The geometry is Euclidean and the angles in a triangle sum to 180°.

The role of adjudicator of which of these solutions represents physical reality rests with the observational astronomer. There are several approaches that have been applied to this problem. One involves a study of the inhomogeneities in the Cosmic Microwave Background, which are visible in Fig. 10.9. These cosmic microwave temperature fluctuations are believed to have been imprinted shortly after the Big Bang, and describe fluctuations in the density of matter in the early universe. They have been frozen in time as the universe transitioned from a plasma to a transparent gas of atoms during the Recombination Era, 380,000 years after the Big Bang.

An analysis of the Cosmic Microwave Background, similar to the frequency analysis that is routinely performed to determine the harmonic structure of ordinary sound waves in air, reveals the fluctuations present at the time of the Recombination Era. (See Appendix 11.1 for more details of this procedure.) The measurements of the CMB shown in Fig. 10.9 were undertaken by the Wilkinson Microwave Anisotropy Probe (WMAP), launched by NASA in June, 2001, and operating until 2010. The Planck space observatory, operated by the European Space Agency from 2009 to 2013, improved on the observations made with WMAP. The results obtained from the Planck observatory show that the universe is topologically flat at large scales to within 0.5%. This is an impressive but somewhat strange result that warrants further explanation.

11.6   Dark Energy and the Concordance Model of the Universe

To achieve such a flat universe requires a very precise and critical matter/energy density. This fine tuning is disturbing because it is surprising that any physical quantity would assume a critical value, such that the smallest change will induce the universe either to collapse, or to expand forever at a finite expansion rate. However, throughout the universe there are a substantial number of quantities with values which, if modified even slightly, would change the nature of the universe such that life as we know it would not be possible.

An example is the strength of the strong nuclear interaction. A slight increase (2%) in the strength of this force would have enabled all the hydrogen in the early universe to bind into diprotons, rather than deuterium. (A diproton is a nucleus comprised of two protons, whereas deuterium is a nucleus formed from a proton and a neutron.) This would have drastically altered the physics of stars, where the elements essential to life are forged. One might say that unless these quantities had been adjusted such that human life is possible, we would not be here to observe and measure them. As we have seen earlier in the present Chapter, this concept is known as The Anthropic Principle.

Putting these philosophical speculations aside for the moment, the problem remains that visible matter accounts for only an estimated 5% of the required energy/matter density. Even including the dark matter required to explain the dynamical behaviour of star clusters, galaxies and galaxy clusters, which is estimated to form 24% of the energy density, the total "explained" density is only 29%. That leaves 71% of the energy/matter density in the universe unaccounted for. To add to the puzzle, this missing component must be something that does not gravitate, otherwise the accelerated expansion of the universe observed by astronomers would not be possible. Indeed, the missing component must exert a form of "negative pressure" to induce this acceleration. Since it produces physical effects, without being "matter", this mysterious ingredient has been named "dark energy".

We see therefore that from the end of the 20th Century, the picture of the universe accepted among astronomers has evolved, and converged towards what is now called the Cosmic Concordance Model (a.k.a. the Cosmic Standard Model). It is also known as the ΛCDM model. Here Λ (lambda) is the cosmological constant that Einstein dubbed as the biggest blunder of his life (see Chap. 10). Cosmologists have arbitrarily reintroduced that constant into the equations of General Relativity, and adjusted its value to account for the accelerated expansion of the universe. The mathematics is thus satisfied; however physics is left with the burden of finding a suitable interpretation for Λ.

As we have seen, because of the observed accelerating expansion of the universe, dark energy must function as a type of anti-gravity. We may think of it as being a fluid permeating the universe, but its density always stays the same. Normally we would expect that in an expanding space any fluid progressively dilutes, but this is not the case for dark energy, which dilutes much more slowly than the universe expands. Reintroducing the cosmological constant of Einstein to GR provides the simplest formalism to describe dark energy, since it is constant in both space and time.

The introduction of an ad hoc arbitrary constant into a physical theory is an indicator that one's physical understanding is far from complete. Ideally the value of the constant should be derivable from other principles. However, attempts to derive the value of Λ from quantum field theories have been spectacularly unsuccessful, with the measured value of Λ being only 10^{-120} of the theoretically predicted value. This discrepancy of 120 orders of magnitude has been described as "the worst theoretical prediction in the history of physics" [6]. In summary, we do not know what dark energy is, but it seems to work, and provides a good fit to many cosmological observations. Of course, there is no guarantee that sometime in the future it will not be replaced by an altogether different theory. Such is the nature of science.

Let us turn now to the remaining CDM part of the ΛCDM model. The letters CDM stand for Cold Dark Matter. We have seen in an earlier Section of this Chapter that dark matter was introduced to explain the observed effects of gravity in large-scale clusters of matter. The various hypotheses proposed by theorists about the nature of dark matter may be grouped into three categories: cold, warm and hot dark matter. The three categories are related to the free-streaming length[1] of the particles (or to their speed): cold means slow, i.e. non-relativistic; warm means weakly relativistic and hot means relativistic. None of the three types is expected to interact strongly with ordinary matter, but each has a different influence on the size of the structures (galaxies, galaxy clusters, and superclusters) that are observed in the universe. Of the three types, Cold Dark Matter provides the best description of the distribution and size of observed structures, and as a consequence it has been included in the Concordance Model.

Below, we summarise the evolution of the universe described by the Concordance Model, making use of a graphic provided by NASA (see Fig. 11.4), where space–time appears as a bell-shaped hyper-surface.

In Fig. 11.4, the three dimensional surface is depicted as a two dimensional expanding envelope. This is an example of the usual compromise forced on mapmakers when representing a three dimensional surface (e.g. the earth) on a plane sheet of paper. The third dimension of Fig. 11.4 is time. The universe comes into existence on the left with the sudden appearance of the Big Bang singularity. Immediately, the limitations of such a depiction become apparent when one asks what occurs in the region shown to the left of the singularity. The question cannot be answered. This region of the figure has no meaning because time (as well as space) is created at the instant of the singularity.

Following the Big Bang, and close to it, events are dominated by the interaction of strong gravity with quantum fluctuations. As we have indicated earlier, the laws of physics in this region are not well understood. Then follows an inflationary exponential expansion, and in the minutest fraction of a second, the universe explodes in scale from the microscopic to the cosmic.

With the burgeoning expansion, the temperature of the content of the universe decreases, and the synthesis of nuclear particles begins. The nuclear plasma continues to cool and the expansion rate of space slows down until the Recombination Era is reached at a cosmic age of 380,000 years, and neutral atoms are formed.

---

[1]The free-streaming length is the average distance a particle can travel without interacting with matter scattered along the way. The weaker the coupling with ordinary matter, the longer the free-streaming length.

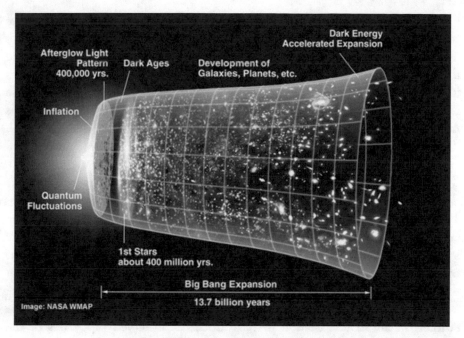

Fig. 11.4 Pictorial view of the evolution of the universe. The bell-shaped surface (which should be three-dimensional) is the border of the visible universe; cosmic time is measured along the surface starting from the initial singularity. Image: public domain, courtesy of NASA WMAP Mission (https://commons.wikimedia.org/wiki/File: CMB_Timeline300_no_WMAP.jpg (accessed 2020/6/19))

Following the Recombination Era, there ensues a "dark age", during which atoms start to coalesce around positive fluctuations of the matter density, thereby increasing the local temperature and pressure. There are no visible light sources during this period, whence the name: Dark Age. The darkness is the reason our telescopes can probe through to the Recombination Era. Finally the temperature and pressure of the agglomerates increase sufficiently to ignite nuclear fusion, and the first stars are born. Once again light is emitted in the universe. Gravitational attraction groups the stars into galaxies, clusters, superclusters, and filaments, etc.

Up to this moment of time, the matter/energy density has been sufficient to allow gravity to slow the rate of expansion. However, the expansion progressively dilutes the matter density, thereby reducing this braking effect of gravity. Dark energy, which as we have noted above, does not become diluted with expansion, begins to dominate the process and convert the expansion from a decelerated to an accelerated stage. This is the phase in which we are now living. In Fig. 11.4, transverse sections of the bell are shown as circles, but in the real world they are spheres, and each of them corresponds to the frontier of the visible universe.

The Concordance Model is currently the most popular model of the universe among cosmologists. Its strength is in its relative simplicity and internal consistency. However, the fact that in this model 95% of the universe is "dark" (or unknown) presents a serious challenge to its credibility, and it is not without its doubters, as we shall see in the next Section.

11.7 Alternative Cosmological Models

The conceptual difficulties with Dark Energy that we have outlined in the last Section have spawned a vast literature describing attempts to find more intuitive or better founded solutions to the problem. It is not within the scope of this book to cover these endeavours in detail. However, some indication of what is involved is appropriate.

Many theories begin by introducing a field, or fluid, whose properties are tailored in an ad hoc fashion to reproduce the observed properties of the universe. The names used for these quantities are sometimes whimsical, such as "ghost energy" or "quintessence". The latter is a reference to Aristotle's "fifth essence", an ether (aka aether) that was supposed to pervade the skies.

A common approach in physics is to draw an analogy between one area of physics and another, in the hope that the first will throw light on the second. For instance, in nuclear physics it is of use in some instances to compare the nucleus with a drop of liquid, and assign to it a "density" and "surface tension". In cosmology, dark energy is sometimes treated as arising from the deformation of a four dimensional elastic continuum. This model is the Strained State Cosmology. For an elastic material in three dimensional space, the strain energy is related to the resistance of the material against an imposed deformation. The accumulation of strain energy hinders the deformation and forces the material to return to the unstrained state. In a four dimensional analogue, if we regard space–time as not just a mathematical artifice, but an entity with physical properties of its own, a strain energy density can be associated with the presence of curvature, and drives an accelerated expansion towards an asymptotic unstrained and completely flat state. It is the analogue of a twisted piece of elastic material returning to its flat state. This theory succeeds in explaining the observed accelerated expansion and also the exponential expansion close to the origin.

Another class of theories involves modifications to Einstein's General Relativity, or even its replacement. However, at this moment, the ΛCDM model remains the conceptual framework that is used by cosmologists as the basis for interpreting observations, even though it contains ingredients that are counter-intuitive and difficult to comprehend.

So now we have arrived at the end of our journey through space–time, and find that in spite of the spectacular progress achieved, we should adopt a humble attitude with respect to our accumulated knowledge. When we look up at a clear night sky and behold the vast expanse of stars stretching across the heavens, we surely feel the same sense of awe and insignificance as our early ancestors. In the intervening millennia, we have learned much, but there is still much to learn. Not all our questions have been satisfactorily answered, and the future will surely show some of our current answers to be wrong. Science, by its nature, is evolutionary, and *Homo sapiens* is a creature driven by curiosity. Today's physicists and cosmologists have left an abundance of puzzles to occupy the fresh minds and curiosity of succeeding generations.

References

1. Moffat JW (1993) Superluminary Universe: a possible solution to the initial value problem in cosmology. Int J Mod Phys D 02(03):351–365. https://doi.org/10.1142/S0218271893000246. Accessed 17 June 2020
2. Oort JH (1932) Bull Astron Inst Neth 6:249
3. Zwicky F (1933) Helv Phys Acta 6:110
4. Milgrom M (1983) A modification of the Newtonian dynamics as a possible alternative to the hidden mass hypothesis. Astrophys J 270:365–370
5. Epps SD , Hudson MJ (2017) The weak-lensing masses of filaments between luminous red galaxies. Mon Not R Astron Soc 468(3):2605–2613. https://doi.org/10.1093/mnras/stx517. Accessed 18 June 2020
6. Hobson MP, Efstathiou GP, Lasenby AN (2006) General relativity: an introduction for physicists. Cambridge University Press, Cambridge, p 87

Part III

Open Questions

12

Issues for the Future

12.1 Doppelbelcher and the OOO

Felix Doppelbelcher Jr was a very disgruntled young man. He had just learned that his stipendium at the prestigious Fachgeplonk University in Omsk had been terminated. Not many in the general public had ever heard of this Institute, or indeed of Omsk for that matter, so he knew his termination would cause little stir in the world at large. However, within the small circle of Omphaloskepsis graduates, or OOO (i.e. the Omphaloskeptics of Omsk), as they liked to call themselves, his expulsion was a major talking point. (Omphaloskepsis, a widely pursued activity in many academic circles, is usually known by its alternative name of "navel-gazing". Only at Fachgeplonk University is it recognised with its own Faculty.)

The aim of the OOO was to return physics to the embrace of philosophy, from which they believed it had unwisely departed, shortly after the debunking by Galileo of so many of Aristotle's assertions. The normal OOO modus operandi therefore involved the execution of thought experiments, and the development of mathematical models of imposing complexity. In this regard, the Dean was typical of most old-school OOOs, who regarded any knowledge obtained by experimentation as somehow of a lesser worth than that derived from purely cerebral processes. Doppelbelcher, however, was a reactionary who maintained that the

OOO would never be accepted as legitimate physicists until they tested their conclusions in the laboratory. He had been in the process of carrying out such an experiment, involving a box, a radioactive source, a Geiger counter, a cyanide capsule and the Dean's Persian cat (since deceased) when he was discovered by the Dean. This misadventure had led to his expulsion.

The real reason for his demise, however—or so Doppelbelcher avowed—lay with his assertion that the fundamental constants of nature, in particular Planck's constant h and the speed of light c, were unstable. Why was this proposition so unpopular? Surely it was only what would be expected from Quantum Mechanics, where atoms could jump spontaneously from one energy level to another. Why should one expect the fundamental constants to be exempt from similar erratic behaviour? Clearly, they should not. However, once ideas become encapsulated in the credos of physics, it is very difficult to overthrow them. In this case, he knew the Dean was saving himself from a long, animated debate over a philosophical principle by ousting him on the pretext of a Schrödinger's Cat experiment gone wrong.

Doppelbelcher could not understand why his ideas had received so little support from his peers. Evidence of sudden changes of state abounded when one knew where to look for them. Who has not lost their keys, only to have them turn up a few seconds later in some unlikely location? Who has never felt someone behind their back staring at them, and yet, when they turned quickly to catch them out, they had vanished? It could all be explained by a sudden fluctuation in Planck's constant.

Try telling that to the OOO, thought Felix. "You're forgetting Occam's Razor," they preach. "Your ideas are too complicated. Remember, the simplest explanation is always the best." This from a mob who believe we live in an eleven dimensional universe—one of an infinite number of parallel universes that we can never visit—and that ninety-five percent of its contents are invisible. "Lord, give us a break!" he muttered aloud.

"Well, they are not going to win," said Felix, straightening his jaw. "Data must always triumph in the end." His eyes shone with an awakening passion. "And why should I care about some piddling stipendium when my rich aunt's favourite charity just happens to be the Omsk Home for Stray Cats." He grinned at the possibilities presented by such a handy source of raw material. You never know, after he'd finished with the Schrödinger investigations, there might even be a few animals left over to pursue the occasional diffraction experiment.

* * *

(**Editor's Note**: We are relieved to report that Doppelbelcher's casual attitude to the ill-treatment of cats is not shared by any of his fellow physicists, and eventually led to his internment in a Siberian salt mine, with only his radioactive source, Geiger counter, and cyanide capsule for company. There he remained for many years, until one morning his cell was opened and discovered to be empty. Speculation still rages over his mysterious disappearance, with the suspicion that some spooky quantum phenomenon has been at play.)

12.2 Time to Jazz It Up

We have now reached the final Chapter of our book, and it is therefore appropriate that we round off our journey with a tantalising glimpse for the reader into some of the areas of physics where new discoveries are being made, and new ideas put forward, almost on a weekly basis. As is clear from our tongue-in-cheek introductory story above, there is often a multiplicity of views in these largely unsettled fields, for physicists after all are only people, with all their human frailties.

In his entertaining and informative book, "*The Jazz of Physics*" [1], cosmologist and jazz musician, Stephon Alexander, explores the connection between music and physics, a relationship that can be traced back to the Greek mathematician, Pythagoras. He also discusses the importance of beauty in physics, a concept that we first encountered in Chap. 3.

In the early stages of his career, while a graduate student at Brown University, Alexander asked one of his mentors, Robert Brandenberger, what he thought was the most important question in cosmology. Our readers will by now no doubt have some answers of their own to this question, e.g. "what caused the Big Bang?" or "how do we resolve the conflicts between quantum mechanics and relativity?" After pondering for some time, Brandenberger replied: "*How did the large scale structure in the universe emerge and evolve?*"

The more one considers this response, the more profound it appears. If, as is now generally accepted, the universe began with a large cataclysmic explosion, presumably from an incalculable amount of energy with no inherent order, what are the origins of the order that now exists, notably clusters of galaxies, individual galaxies, stars, solar systems, and ultimately life? After all, "*nothing will come of nothing,*" [2] as the Bard so rightly, if somewhat plagiaristically,[1] informs us.

The word "evolve" in Brandenberger's reply is evocative. A human develops from fertilised egg to adult because of the presence of DNA in the nucleus of the egg cell. Is there some equivalent package of information present at the initial Big Bang singularity that foreordains the evolutionary path, at least in broad principle, that the universe must take? In his theory of the evolution of species, Charles Darwin identified *natural selection* as the driver that guides the evolution of a species and creates order from the chaos of random muta- tions. Is there a corresponding driver, perhaps involving gravity, that produces order out of the early chaos of the universe?

On the other hand, what appears to be order may actually result from something very simple. One of the authors is reminded of a fishing excursion some years ago, when a clumsy comrade knocked over the bait container, depositing hundreds of writhing maggots in a heap on the carpeted floor of our boat. As we watched, we saw an almost perfect, rapidly expanding, circle of larvae advancing with military precision outwards from where the heap had landed. They were not, of course, following the command of some insect drill sergeant, but had simply headed off in the direction they happened to be pointing when they alighted. The random nature of the fall ensured that all directions were equally covered, their speeds were approximately the same, and the result was the circle we observed. There are many examples of complex behaviour arising from a few underlying basic rules in social insects (bees, ants, etc.), flocks of birds, schools of fish [3], and even crowds of humans. The science of thermodynamics, including concepts such as entropy, has been successfully applied to interpret crowd behaviour.

So perhaps the emergence of structure in the early universe is just chance, and we are simply fortunate that our world seems to be tuned to support the development of life? If so, we have indeed won the lottery, because a slight change in any of a number of fundamental physical constants, or a small variation in the physical laws or in the universe's conditions in its early stages, would have produced a universe where life as we know it would be impossible. We shall discuss the origin of the universe further in Sect. 12.4. It is also possible that our universe is not unique, and we are surrounded by innumerable other inaccessible universes, in which case we are only aware of our own universe precisely because it is one in which life is possible.

---

[1]The bard was paraphrasing the much older: *Nothing can be made from nothing—once we see that's so, already we are on the way to what we want to know.*—Lucretius, *De Rerum Natura*, 1.148–156.

Some readers would say that such questions belong in the fields of philosophy, metaphysics and/or religion, because it is impossible to ever explore these issues with the methodology of science, as no observations of the early days of the universe are possible. Since the time of Galileo, theory and experiment (or observation) have progressed together in a symbiotic relationship that has produced our modern technological world. However, there are many who believe almost religiously (as did Dirac) that beauty is a better guide to the veracity of a theory than experiment or observation. In part, this attitude has been forced upon cosmologists by the difficulty (or even impossibility) of obtaining observations in the regions of space-time now of critical interest (e.g. near the big bang singularity).

Fundamental Particle physicists are also faced with difficulties, albeit of a somewhat different nature. In their case, following the triumphant discovery of the Higgs boson in 2012, the pickings from Cern's Large Hadron Collider have been meagre indeed. Most of the particles predicted by the various extant theories have steadfastly refused to make an appearance, and theorists have attributed this reluctance to insufficient energies in the collider beam. They have registered their desire for a more energetic accelerator, but such devices, with their billion dollar price tags, must compete in priority with feeding the starving masses, solving the climate change problem, curing cancer, and other more urgent issues.

As a consequence of the difficulty in obtaining appropriate data in the two fields that lay at the vanguard of modern physics, we now appear to have entered an era of *post-empiricism*, where the beauty of the mathematical equations is used in selecting which areas to pursue. But beauty according to whom? Without the constraining hand of experiment to guide us, beauty is now given a license it has never had since the time of Galileo. The lauding of beauty (aka symmetry) in the Twentieth Century resulted in some stunning successes, for instance, in Quantum Electrodynamics and the Standard Model of Particle Physics. However, if a theory did not pass the test of experiment, beauty was not of itself a sufficient criterion for its acceptance.

For many astronomers, the Steady State Theory of the universe was more "beautiful" than the currently accepted theory, with its ugly Big Bang singularity at the origin of space-time. However, the Steady State Theory was discarded without compunction when sufficient observational evidence had accumulated against it. To further complicate the issue, beauty is as subjective in mathematical equations as it is in art galleries. Some researchers show a very human tendency to find beauty in their own equations, and ugliness in those of their colleagues. In any case, as Sabine Hossenfelder asks in *Lost in Math*: "*Why should the laws of nature care what I find beautiful* [4]?"

We began our journey in this book with a look in Part 1 at the earliest attempts of humans to make sense of their world. We saw how progress was erratic until the development by Galileo of what is now known as the scientific method, whereupon began the bourgeoning of science that has impacted the lives of everyone on our planet. Now it seems that some domains of physics are beyond the reach of experiment or observation, and thus beyond the ken of what we mean by science. Perhaps we should just accept this, albeit with a degree of despondency. However, human curiosity being what it is, it is difficult to prevent ourselves from venturing into regions that are customarily regarded as metaphysics.

12.3 The Beginning of Time

Surely every intelligent being has at some time speculated about the origin of their universe. However, arguably, nobody before Galileo had ever really doubted the infinite range of time, and even believers in the creation of the world by a God (or some other more or less defined entity), believed that, if not the world, then at least God had always existed. Likewise space was thought to range unbounded in its three classical dimensions.

Today, the current consensus is that time originated with the "Big Bang", i.e. it does not run back forever (to minus infinity), but starts at a given singularity. (We must be careful not to say at *a given point in time*.) In other words, when the universe was born, space and time were born together in the composite known as space-time.

As a consequence, to use a simpleminded analogy, the universe is like a movie that has been switched on and lasts as long as the director chooses to allow it. However a movie runs for only a limited time, while our universe might last forever, with an expansion leading to an extremely vast and cold desert. Or it might terminate at any time in a collapse, or unimaginable cataclysm, in which case the whole process might begin anew. We hinted at this possibility with our analogy of the *ouroboros* in Chap. 3.

In the case of a movie, we can learn who the director was, where it was made and why, how much it cost, and a host of other details. This information is available because the movie was produced in the framework of time. There is a "before", which can tell us how the director conceived the idea, and how he or she obtained the requisite financial support. There is a "during" with all the paraphernalia that creating a movie involves, and there is an "after", in which the audience can enjoy (or hate) seeing it, and learn of its eventual success. In our Universe, there is no "before", because time itself had no existence "before". We cannot ask what its cause was, because a relationship of cause-effect requires time in order to exist.

At least this was the picture of the universe held by cosmologists since Einstein staggered the world by overturning Newton's theory of gravity. Around the turn of the Twenty-first Century, however, there were new proposals put forward that modify our conception of the earliest moments of the universe. They are controversial, and certainly not accepted by all, but have in recent years garnered increasing currency. We present some of the basic concepts here.

In Chap. 10, we explored the beginning of the universe as far back as we can reasonably go. As we emphasised there, the first 380,000 years of our world's existence remains shielded from our view, due to the impenetrability of this region to electromagnetic radiation. The newly discovered gravitational waves are currently the only foreseeable possibility of direct observation of this period in the future.

Although direct observation is not possible, there are residual effects from this period that linger into the later development of the universe, and provide us with some idea of what may have gone on before. Indeed, we find ourselves very much in the same position as Plato's slaves (see Chap. 1), whose observations of the shadows on the cave wall were all they had to provide them with an inkling of the reality in which they were immersed. An example of these residual effects is the incredible isotropy of the Cosmic Microwave Background, and the nature of the few remaining inhomogeneities. As we saw in Chap. 10, a period of very rapid inflation, lasting from 10^{-36} s to 10^{-32} s after the Big Bang, has been widely accepted as an explanation of these effects.

Nevertheless, there will always remain a desire to explain the inexplicable. Our universe appears to be tailored for the evolution of life. This is one of the outstanding enigmas facing us as intelligent humans. It may be of little importance in our daily lives. It has no impact on climate change, the international economy, or who will win the hundred-metre sprints at the next Olympics. It is just a question that nags at us, and refuses to be put aside.

Naturally, there are theories that have been constructed to address this issue. They also attempt to explain the preponderance of matter, compared with anti-matter, in our universe; the laws of physics; and the possible unification of the four fundamental physical forces. Most of these theories involve a multiverse, in one form or another. We shall discuss the multiverse, which is a whole swathe of parallel universes, in Sect. 12.7. For the moment, let us discuss the *very* early universe.

So enrapt were cosmologists by the success of the inflation model of Alan Guth (see Chap. 10) that they applied it to the era immediately after the Big Bang. At these times, because of the infinitesimal size of the space containing all matter/energy, gravity is expected to dominate all physical interactions. Indeed, all of the forces of physics may have been unified at this time. Unlike in Guth's original proposal for inflation, it has been suggested in a new model known as "*eternal inflation*" that inflation persists forever. Within the inflation field, quantum fluctuations occur spontaneously and randomly from place to place and time to time. Some of these fluctuations result in regions where inflation stops, and a universe such as ours begins evolving. In others, inflation continues rapidly. In this theory, our universe is a single "pocket" or "bubble" embedded in a vast assemblage of universes in an inflating space without end.

Figure 12.1 provides us with a picture of what such a universe might look like. Our universe is one of a myriad of bubbles. Each bubble is a different universe with different values of the fundamental constants, some contain predominantly matter, some predominantly antimatter, and even the laws of physics may vary from one to the other. We live in a universe suitable for life (probably one of very few) because if we didn't, we wouldn't be around to

Fig. 12.1 Soap bubbles provide a model for bubble universes. We live in a "bubble", surrounded by countless other bubbles. Each has its own physical laws, and fundamental constants. Most of them are unsuitable for life

report on it. The model provides an answer to the question: what happens before the Big Bang? But leads to another question, what happens before t = 0, and the commencement of "eternal inflation"?

12.4 Interfaces between Theories

It is at the interface between theories where some of the most challenging areas of physics lie. We have seen, for instance, how, as the number of molecules interacting with each other in a gas increases, physicists were forced to resort to a statistical approach to the problem. It is not that they believed that the laws of mechanics did not hold, but only that they could not be solved for a system containing so many molecules. This led to the development of thermodynamics, which in itself has been applied to areas way outside of physics, such as the description of crowd behaviour at rock concerts.

The success of Quantum Mechanics and Relativity in the 20th Century, coupled with the continuing dominance of Newtonian Mechanics for the solution of the physics problems of everyday life (building bridges, cars, rocket ships, etc.) has led us to several important new interfaces. These are discussed below.

Newton meets Einstein

The interface between classical mechanics and relativity presents comparatively few problems. The Theory of Special Relativity could be applied in Newton's domain, where the velocities of objects are small compared with the velocity of light, as it is accurate there. However, it is too difficult to use in most cases, so we prefer to stick with Newton's formulation. This example is the ideal case of a smooth transition between two different theories operating in overlapping regions, where the classical theory can be regarded as a high-accuracy approximation of the other.

Similarly, the field equations of Einstein's theory of General Relativity reduce to Newton's theory of gravity when the space-time curvature is relatively small (i.e. the gravitational fields are "weak") and the speed of the objects is much less than that of light. Except for the fact that Einstein's field equations are horrendously difficult to solve, we would therefore be able to discard Newton's theory completely. Certainly some of the concepts in the theories of relativity (e.g. mass-energy equivalence, and the dependence of simultaneity, time dilation and the contraction of length on the motion of the observer) stretch our credulity, but with an open mind and a faith in classical logic and experiment, they can be accommodated.

Newton meets Schrödinger and Heisenberg

On the other hand, the concepts of QM are so strange that even a genius such as Einstein could not accept them. Why do we see so little evidence of these weird quantum phenomena, except in the world of atoms and fundamental particles?

Since its earliest days, it has been realised that QM should transition smoothly into classical mechanics as one moves from the microscopic world of fundamental particles and atoms to the macroscopic world of everyday objects. This was recognised by Nils Bohr and is given the name of *Bohr's Correspondence Principle*.

Bohr showed how for macroscopic systems, the energies involved in the interactions of their components are much larger than the energies of individual quantum transitions, and in this case the quantum results reduced to those expected from classical mechanics. In addition, macroscopic bodies are comprised of many smaller particles interacting with each other. The wave functions of these smaller particles are continually being jumbled together as the particles move about and swap energies. The resultant composite wave function for the macroscopic body is therefore "decoherent", which means that it does not produce interference effects. When we walk through a narrow doorway, we have no chance of being diffracted away from our normal path, as a quantum particle would be, passing through a narrow slit.

This all sounds well and good. However, not everybody at the time was convinced by Bohr's arguments, and the controversy is still alive. There are some simple quantum systems, e.g. a particle confined in a box, that do not pass smoothly from a quantum mechanical to a classical description as the energy of the particle increases [5]. Nobel Laureates, Albert Einstein and Max Born, disagreed strongly on this issue, leading to Born dismissing Einstein's obduracy to be the result of his old age, a blow below the belt if ever there was one.

As we have seen in Chap. 5, Schrödinger's qualms about the QM-Classical interface led him to formulate the Schrödinger's Cat thought experiment to emphasise the problem. It may seem rather ridiculous to consider a cat as being in a superposition of states, one in which it is alive and one in which it is dead. However, what if the cat is replaced by a virus? These are considered by many to be living creatures.

An experiment [6] has been proposed to trap a virus in a vacuum, then slow down the virus's movement until it resides in its lowest possible energy state. A laser can then be used to target the virus with a single photon and excite it into a superposition of two states, one where it is moving, and one where it is not. The philosophical implications of an experiment such as this are far-reaching, particularly if it can be applied to even larger living organisms, such as bacteria and tardigrades (water-bears). The authors of the proposal suggest that such an experiment will be a starting point for addressing the role of life and consciousness in quantum mechanics.

Einstein meets with Schrödinger and Heisenberg, and all parties agree to disagree

The remaining interface is where small particles are moving at high speeds in strong gravitational fields. We simply do not know what physics to apply in this region. It is truly a domain where in medieval times a "*here be lions*" sign would have been erected to warn brave souls venturing into this territory that they do so at their peril. Unfortunately however, it is not a region that can simply be dismissed as unimportant, for somewhere within its boundaries lies the origin of the universe.

12.5 Entanglement and Quantum Computing (QC)

The concept of Quantum Entanglement (QE) has already been discussed in Chap. 5, and is surely one of the most intriguing results of QM, with a weirdness that mystified even Einstein. Hence, some readers may wonder why we return to it here in Part 3, where our concern is with future developments of our chosen topic (i.e. physics). The reason is that in the case of QE, a clear demarcation between past and future research, i.e. between what has already been acquired as common wisdom and what can more suitably be defined as current research, cannot easily be defined.

In fact, even after many decades, QE still lies at the frontier of research for several reasons. First, new experiments are currently being carried out with the purpose of not only demonstrating the effect, but of visualising it. A paper [7], published by a team of physicists from the University of Glasgow, led by Dr. Paul-Antoine Moreau, describes an imaging process that enables the phases of both a photon and its entangled *twin* to be recorded simultaneously, thus displaying in a most convincing way the reality of the entanglement.

Second, QE, far from being merely an intellectual curiosity, or even a fundamental theoretical step towards our understanding of the mysteries of nature, may also provide us with an ideal tool for the transition to the quantum world, to which we increasingly appear to belong. Even though at present it may seem farfetched, it seems likely that almost all branches of science will eventually need to include quantum effects for particular applications.

As an example, let us consider chemistry, a field that traditionally was considered something of an "art", rather than a science, although, of course, its methodology is fully scientific. QM, with its strict mathematical basis, has changed all that. When it was first formulated, and for a few decades thereafter, the pinnacle of glory of QM was the quantitative explanation of the hydrogen atom as a composite of two particles: a proton and an electron. Nowadays, QM is applied to complex atoms and molecules, using the most refined computer techniques. Indeed, Quantum Chemistry has become one of the most heavily mathematised of all the sciences.

Let us now turn our attention to biology and related sciences. Just a few decades ago, nobody could have anticipated (except perhaps visionaries like Schrödinger [8]) that a new science, Quantum Biology, would become a major field of research. Although still in its infancy, Quantum Biology claims to be able to explain hitherto unexplained puzzles, such as the amazing navigational skills of migratory birds in their transcontinental flights [9], or the extraordinary efficiency of the photosynthesis process [10].

Based on quantum entanglement and on the superposition principle, i.e. on the idea that, as we saw in Chap. 5, a quantum system is in all of its possible states at the same time, until it is measured, Quantum Biology offers explanations that appear more convincing than their classical counterparts. However, there are still many open questions, as one might expect for a topic that is the subject of active research.

In spite of the extraordinary relevance of Quantum Biology, the most intriguing and sought after application of Quantum Entanglement (and of the Superposition Principle) is Quantum Computing (QC), to which we devote the remainder of this Section. It is interesting that the concept of a wave function associated with each particle, which for physicists has been one of the most difficult puzzles and headaches for decades, has now become a blessing in disguise, from the point of view of QC. In fact, a *classical* particle can carry very little information, i.e. its space-time coordinates and momentum, while the wave function of a *quantum* particle consists of an array of indefinitely many numbers.

Despite Einstein's misgivings, quantum entanglement is now being incorporated into innovative QC technologies that promise (or threaten) to extend the power of modern computing systems by a factor of up to 100 million. Instead of the conventional bit, with values of 0 or 1, which is the basis for arithmetic in the desktop (or Turing) computer, a QC uses a *qubit*, or quantum bit. With a qubit, because of its quantum nature, it exists in a superposition of states, and thus can have the value of 0, 1 and everything in between at the same time.

Quantum Entanglement is employed to make measurements of the qubit's state indirectly, to preserve its integrity. If an outside force is applied to two atoms, it can cause them to become entangled. The instant one of the atoms is disturbed, its wave function collapses and it chooses one spin, or one other quantum value; simultaneously, the second (entangled) atom will choose an opposite spin, or value. This technique allows qubits to be interrogated for their value without actually observing them.

Recent advances have shown that quantum computers are on the verge of solving problems that are too hard for classical computers to tackle. This is called the moment of *quantum supremacy*. However, there are still many problems to be addressed, in particular, reliability. Any small perturbation can cause the qubits to lose entanglement, producing wrong computational results: one might easily argue *"what is the use of a superfast computer if it delivers the wrong answer?"* Also, if no conventional computer is capable of checking the numerical accuracy of the computation, how do we ever know whether the solution is correct? These issues will undoubtedly be addressed satisfactorily in the not too distant future.

Some of the main areas of application of QC technology will be in domains that are themselves quantum, such as the determination of the structure of atoms and molecules, and the design of new drugs to cure diseases. There are likely to be applications in Artificial Intelligence, such as quantum neural networks, and in the deciphering of encrypted information. They may well render obsolete the encryption currently employed by banks and others for secure on-line communications. There are thus many implications for financial and national security. As it becomes ever clearer that we live in a quantum world, the range of application of quantum computers will continue to grow, and they are likely to assume increasing importance in guiding our future choices.

12.6 Towards a New Copernican Revolution

Before we open discussion on specific issues in modern physics that merit further examination, it is timely to revisit some of the questions raised in Part 1 that influence the methodology employed by physicists in their pursuits. In particular, in Chaps. 1 and 2 we discussed the nature of reality, what we mean by understanding, and the role played by logic in mathematics, and as a consequence, physics.

Central to the early years of *Homo sapiens* on this planet was an underlying assumption of human intellectual superiority. So prevalent was this attitude that a name was coined to describe it: *anthropocentrism*. Initially, innate human intelligence was used to improve the chances of survival and comfort of family and friends. However, after having reached, in the course of tens of millennia, a certain level of prosperity and security, humans began to gloat on their superiority, to the point of not accepting that they were just another animal species. Since no other anatomical differences could be found to explain their perceived nobler role, they thought that the human brain must possess some discriminating characteristic, which they called *reason*. Whether God given, or in some other manner inserted into the human brain, and being common to all, irrespective of their life experiences, *reason* had to be "*a priori*", or so they thought, compared with the "*a posteriori*" of external phenomenology.

Perhaps the latter point deserves some further clarification. If we live in a particular part of the earth, and nobody is around to tell us about events from outside our range of experience, we must derive our knowledge solely from (or *after*—that is the meaning of *a posteriori*) our experiences of our external world. However, in the case of arithmetic, our logic and intuition lead us to the same results as those obtained by other people, irrespective of where they live, which language they speak, or what has been their life experience. Hence reason, which enables us to perform these arithmetical operations (and many other logical deductions) must come from within us, *before* our experience of the surrounding world (i.e. *a priori*).

The perceived privileged role of humanity also led to another important consequence. Since *Homo sapiens* (whether created by God or not) were of such extraordinary significance, the earth on which they lived had to be the centre of the world, a fact confirmed by their observation of the sun, which was seen as rotating around the earth, and of all the stars, which were apparently nestled in a sphere with the earth at its centre. In short, humanity appeared to be the *raison d'être* for the whole universe.

However, to the chagrin of generations of philosophers and pious believers in religion of one form or another, along came Copernicus and Darwin. They shook this long-established vision, swollen with its hubris and self-indulgence, to its foundations. Darwin was ridiculed by the establishment for his assertions on the evolution of humanity and apes from common ancestors, as the cartoon in Fig. 12.2 demonstrates.

We now know that we live in the periphery of one of billions of galaxies, and that our sun is just an average star among trillions, and our planet one among billions of others, some of which may have a biosphere comparable with, or even richer than, ours. In Chap. 10, we saw in Fig. 10.7 a small section of space containing thousands of galaxies, each of which is itself comprised of billions of stars. Images such as this one underline the preposterous nature of the pre-Copernican view of the world.

Fig. 12.2 "A Venerable Orang-outang", a caricature of Charles Darwin as an ape published in *The Hornet*, a satirical magazine, in March 1871. Image public domain due to age (https://commons.wikimedia.org/wiki/File:Editorial_cartoon_depicting_Cha rles_Darwin_as_an_ape_(1871).jpg (accessed 2020/6/21))

Today, we still may not know whether there are intelligent forms of life on other exoplanets,[2] but we do know that here on earth we are the result of an evolutionary process that has led us inexorably to our current position as the most intelligent of earth's animals.

In Chap. 2, we have discussed the nature and origin of classical logic. This way of thinking, as encapsulated in mathematics, is at the very heart of science, and thus responsible for most of our technological development. We saw in Chap. 2 that both the platonic (aka *a priori*) and empiricist (aka *a posteriori*) viewpoints have their followers. In the platonic view, mathematics is "out there", describing the physical world, and just waiting to be discovered by humans, or presumably other intelligent life forms. In the empiricist view, mathematics is a human invention like any other, which may or may not describe the physical world. It would probably be true to say that most mathematicians would favour the former view over the latter.

In Chap. 4, we saw how Gödel's incompleteness theorems whipped the rug out from under rigorous mathematicians in their attempts to construct all of mathematics by logical deduction from a limited number of basic axioms. Essentially such an approach results in a system that either has contradictions, or truths that can never be proved. As physics is based on mathematics, we might expect the same arguments to apply, i.e. there will be some physical "truths" that can never be derived from theory. This was the basis of Stephen Hawking's change of heart about the unlikelihood of a "theory of everything" that we described in Chap. 4, and follows directly from the platonic viewpoint.

We see from the above paragraph that a change of perception of the nature of reason (from *a priori* to *a posteriori*) is by no means just an abstruse academic game, but has very important ramifications. If the platonic point of view applies, then Gödel demands that physics must follow mathematics, and physicists should explore which "truths" can never be explained in our current view of the universe.

We saw an example of this type of investigation in Chap. 4, where Toby Cubitt, a quantum-information theorist at University College, London, and co-workers, found that the problem of determining whether there is an energy gap between the lowest energy levels in a material is undecidable. This team also wants to study a related important problem in particle physics, exploring why the particles that carry the weak and strong nuclear forces have rest mass, while photons have no rest mass. This question may also fall into the "undecidable" class.

---

[2]Exoplanets are planets that orbit a sun in another solar system, and not our sun.

On the other hand, if the empiricist view holds, our classical logic and mathematics are inadequate to explain all of physics, and perhaps we should be seeking to develop a different logic for this purpose.

It is not surprising that our classical logic works so well in the macro world of classical physics. This is after all the world in which we, and our intelligence, have evolved, and for which classical logic is tuned. No hominid or primate has ever played with toys of molecular dimensions, nor travelled at speeds close to the velocity of light. Had our ancestors been exposed for many thousands of years to the wonderfully strange world of *Alice in Quantumland* [11], the logic that we now use would arguably be different.

It is a very surprising thing that mathematics describes the physical world so well. Wigner [12] and Manning [13] in two separate thought-provoking papers have explored this enigma under the title: *The Unreasonable Effectiveness of Mathematics*. Classical mathematics, based on long chains of logical reasoning from a few basic assumptions (axioms), is routinely used to explain observed physical phenomena. It is also used by engineers to construct bridges, aircraft and other complicated structures, with full confidence that they will function as planned. Symmetries in mathematical equations have led to the prediction of new (and at the time yet unobserved) physics, e.g. radio waves, inferred from the equations of Maxwell (see Chap. 5), and antimatter, from Dirac's equation (see Chap. 8).

Despite the astounding success of classical mathematics in so many areas of physics, there are cases where it falls surprisingly short. Newton's law of gravity was one of the spectacular early successes of physics. It explains the orbiting of one celestial body about another with great precision; however, the addition of an extra body to the celestial system renders the problem insoluble using classical mathematics, and its solution requires an approximation method.

When we realise that most of the world comprises many bodies, both small and large, interacting with each other, the limitations of the bottom-up approach become clear. When the number of interacting bodies becomes immense, physicists have devised statistical methods (statistical physics and thermodynamics) to successfully describe this domain. However, that still leaves an intermediate region where solutions to important problems are difficult to obtain.

In Chap. 2, we discussed alternative approaches to the traditional bottom-up approach to the laws of physics, involving statistical and pattern recognition methods. Recently, an approach to solving the three-body problem using neural networks has been investigated [14]. Neural networks are computers constructed from parallel arrays of processors. They are not programmed in the same manner as a normal computer, but are trained by presenting them with the output of thousands of calculations carried out with traditional methods; they learn by experience, in the same way that the human brain learns to distinguish the image of a tree from that of an egg. They are then able to use this training to solve different, but similar, astronomical problems much faster than is possible with a conventional computer. Inherent in this approach is the difficulty of checking whether the results obtained in this manner are correct. As we saw in the last Section, this problem will be exacerbated in the near future when quantum computers, with their much faster speeds, become available.

In attempting to obtain a reason why mathematics is so successful in theoretical physics, Hamming and later Abbott [15], point their fingers at humanity. We tend to be selective in which problems we address with mathematics; e.g. there is no mathematical theory of truth, beauty or justice. When existing mathematics cannot be applied to a particular physical problem, we try to invent new forms to use. We have already described in Chap. 4 the extension of our numbering system from simple integers to complex numbers as the centuries passed, and the need to address more complicated scenarios arose. In fact, as we have seen in the last paragraph, there are vast domains of interest where the logical bottom-up approach fails.

Finally, it is suggested that the role of natural selection in the evolution of our species would have favoured those individuals who can follow long chains of close reasoning. We should point out, however, that Hamming is unconvinced on this point.

We have, in earlier Chapters, highlighted the contradictions between the theories of Quantum Mechanics and Relativity, both of which are formulated with classical logic. In particular, Quantum Mechanics is a probabilistic theory, where predictions are of a statistical nature, while Relativity is deterministic, with precise predictions of the trajectories of objects. In Quantum Mechanics, particularly in Quantum Entanglement, the collapse of the wave function fixes the properties of particles simultaneously, while in relativity simultaneity is different for different observers who are moving with respect to each other.

It is entirely possible that scientists will find a way to understand and accept even the strangest quirks of modern science by stretching the limits of classical logic. However, it is also possible that a Copernican revolution in the field of logic, analogous to that which has occurred in our understanding of our place in the cosmos, might turn out to be a more direct route towards a better understanding of those parts of the world that are not our natural domain. Hopefully the new logic, being unhindered by an anthropocentric root, would open us up to a simpler and more natural understanding of modern science and, as a consequence, to new discoveries. Perhaps in the new logic even the contradiction implicit in Goedel's Incompleteness Theorem, which naively appears to be a logical demonstration of its own logical indemonstrability, would disappear. From such a logic we might be able to derive, as a particular case, our conventional and well established logic, in the same manner that Newton's mechanics is a particular case of Einstein's Theory of Relativity.

The first steps towards developing a Quantum Logic were taken in 1936 by Garrett Birkhoff and John Von Neumann [16]. In the intervening decades considerable work has been done by many to explore various non-classical logics. Noted American philosopher, Hilary Putnam, was a strong advocate of a non-classical logic. In 1968 he wrote [17]: *Logic is as empirical as geometry. … We live in a world with a non-classical logic.* However, during his career, he underwent a change of heart, and in 2005, he recanted [18]: (In 1968) *I even proposed an interpretation which involved a non-standard logic, but, as I explained [19] (in 1994), that interpretation collapsed. … More recently, however, it has occurred to me that I should instead attempt to classify all possible interpretations (or all possible interpretations that do not involve giving up classical logic), since I had satisfied myself that the approach via a non-standard logic would not work.*

Whatever the eventual outcome of research in this area, it is certainly a possibility that aliens (if they exist at all) may well think quite differently from us. For instance, perhaps their pattern-recognition capabilities will be far more developed than ours, enabling them to discover in an instant "truths" about their world that our one-step-at-a-time, logical approach fails to reveal. All this is speculation, but it is useful to remind ourselves from time to time of how insignificant we are, and how we have a history of allowing our anthropocentric viewpoint to distort our interpretation of the universe.

12.7 The Multiverse: Physics or Metaphysics?

The idea that there exists a multiverse, or large number of parallel universes, has been around for a long time in many different fields. It is a common theme in the Science Fiction genre, and has been proposed in a number of different areas of physics. Schrödinger, in a 1952 lecture, suggested that the alternative histories implied by his equations in QM might actually be realised in other parallel universes.

The concept of a multiverse in cosmology is still highly controversial. A strong proponent, cosmologist Max Tegmark from the Massachusetts Institute of Technology and the Foundational Questions Institute, believes that the multiverse is not a theory, but rather *is predicted by a theory*, and a widely accepted one at that, namely the Concordance Model describing the inflation of the early universe (see Chaps. 10 and 11). If we are prepared to believe this theory for describing the evolution of the universe after the big bang, we should be prepared to accept its other predictions, unless there is strong evidence to the contrary [20].

Tegmark identifies four different levels of parallel universes within the Multiverse. In a Level I parallel universe, observers experience the same laws of physics that we do, but with different initial conditions, resulting in a different history. In a level II universe, the fundamental physical constants may also be different. This provides an explanation of the Anthropic Principle that we have discussed earlier. A level III parallel universe involves quantum effects, where every time a measurement is made, the universe splits into other divergent universes (see Chap. 5). In a Level IV parallel universe, not just the fundamental constants, but also the laws of physics are different from in ours.

In the absence of direct evidence of the multiverse, proponents are forced to resort to indirect evidence, and this is where most of the controversy and disagreement between different physicists arises. As we discussed in Sect. 12.2, beauty is regarded by some as a sufficient criterion for the acceptance of a physical theory, even in the absence of corroborating observational evidence. Others demand a much higher standard of proof.

Some claim that the refusal to accept a multiverse is another example of *anthropocentrism*, the placing of our universe above all others, akin to the pre-Copernican period when the earth was thought to be the centre of the known world. Their opponents dismiss the multiverse *as the last resort of the desperate atheist* and an attempt to avoid the creationist implications of the anthropic principle: i.e. the choice is between a multiverse and a creator, so take your pick. The explanation it offers for the fine tuning of the fundamental physical constants is probably the strongest argument in favour of the multiverse. Paul Davies has discussed the Anthropic Principle in detail in his book: *The Goldilocks Enigma* [21].

One might conceivably raise Occam's Razor (see Chap. 3) as an objection to the multiverse. Surely postulating an infinite number of parallel universes to explain our own is a flagrant violation of the good monk's guideline. Others counter this assertion with the argument that the multiverse eliminates the necessity to postulate precise values for initial conditions and fundamental constants, and that controversial concepts such as wave function collapse are no longer needed. We therefore arrive at a similar point of contention to the one that we encountered in Chap. 3 when assessing the leprechaun and his acts of magic: what do we count as an assumption when applying Occam's Razor?

The debate on this question is sure to proceed for many years to come, as it revolves around what is acceptable evidence for a scientific theory, which is not something on which there is universal agreement among physicists. Traditionally, successful scientific theories should be testable, i.e. at least, in principle, now or sometime in the future, be capable of being falsified by experiment or observation. The trouble is that the multiverse theory can explain anything at all, and so cannot be falsified. As such, at this time it probably belongs more in the domain of metaphysics than physics. This may not be a life sentence, however. Other physical theories began their existence in this manner, and became accepted later as genuine theories, after emerging technology provided a way of putting them to a practical test.

For an excellent summary of competing views on this topic, see: *Universe or Multiverse?* by Carr and Ellis [22]. In the following Sections we address some less esoteric issues in earth-bound physics, which also warrant further discussion.

12.8 *The Past is Never Dead. It's Not Even Past* [23]

In Chap. 5, we encountered *Wheeler's Delayed Choice Experiment*, where choices made at the time of an observation appear to influence what has already occurred in the past. In this experiment we saw that our decision whether to use a wave detector (e.g. a sheet of photographic plate that allows us to build up an image of interference fringes), or particle detectors (which respond to individual particles) affects the wave/particle nature of light passing through two slits. In the first case, light behaves as a wave and in the second as a stream of photons. This behaviour appears to be influenced by the observer's decision on which detector to use, even if that decision is made after the light has already passed through or by the slits.

It is this apparent *retrocausality* in the behaviour of the light that creates so many philosophical and conceptual difficulties in the interpretation of this experiment. John Wheeler, the proposer of the original thought experiment, had this to say: *There is an inescapable sense in which we, in the here and now … have an inescapable, an irretrievable, an unavoidable influence on what we have the right to say about what we call the past* [24]. As we have seen in Chap. 6, the Theory of Relativity has shown that the order of events in space-time can depend on the state of motion of the observer. Simultaneity is not something that two observers moving with respect to each other need agree on. However, neither relativity nor classical physics allows the order of two events to be interchanged when one is the direct cause of the other. This is the heart of the paradox that Wheeler's experiment reveals.

One explanation of this experiment, but not one that attracts many adherents, is that we live in a deterministic universe, where everything is pre-ordained. An example would be if the universe were a simulation, and we are comprised of bits of software code. This idea was developed in the series of *Matrix* movies [25] that were popular in the early years of this century. In this case, there is nothing random about the choices an observer makes, even if he or she is under the illusion there is. Our actions are already recorded in the book of time, where past and future lose their meaning.

The explanation, now generally accepted is that it is only at the moment of observation that the photon displays either a particle or wave nature. The act of observation collapses the wave function. Before this the light is described as a superposition of states, some where it passes through one slit, some where it passes through the other, and some where it passes through both. Passage through the slits is not in itself an act of observation, and does not result in collapse of the wave function.

When the observation is finally made, either at the photographic plate or at one of the telescopes, one cannot infer from this what state the photon was in at an earlier time, for then it was in a superposition of all possible states. The phenomenon of collapse of the wave function is one of the most contentious aspects of QM. Some maintain that reality does not exist until we measure it, and that if everybody were to shut their eyes at once, the moon would vanish from existence. This is an extreme interpretation. It is more appropriate to say that quanta exist as unique objects which display both particle and wave characteristics, and that these different characteristics are brought to the fore in different experiments.

Experimenters have performed Wheeler's experiment over very large distances and for increasingly massive particles. In 2017 a team of Italian physicists at the Italian Space Agency's Matera Laser Ranging Observatory (MLRO) split light on the ground and sent the two beams to a satellite 3500 km away. There they carried out the "delayed choice" part of the experiment, with results that showed the quanta had maintained their wave/particle duality over the length of the journey [26].

Of course, Wheeler's Experiment can be applied to material particles, as well as photons; QM is quite explicit on this point. In 2015, A.G. Manning and co-workers at the Australian National University carried out the experiment with an ultra-cold Helium atom [27], and observed the same quantum effects that had been observed with photons and smaller particles.

We may ask ourselves just how large can an object be before QM gives way to classical Newtonian physics. A recent experiment has extended this boundary even further with the report of a two-slit interference experiment showing the wave-like quantum behaviour of molecules comprising up to 2000 atoms [28]. The results cannot be explained classically, and are in excellent agreement with the predictions of Quantum Mechanics.

12.9   Are Space and Time chunky?

In Chap. 5 we saw that, from as early as the 5th Century BCE, Philosophers such as Leucippus and Democritus, queried whether matter, which superficially appears continuous, is actually constituted from small, indivisible particles, which they called atoms. This idea lay dormant for two millennia, until revived and confirmed by scientists such as John Dalton, Robert Brown and Albert Einstein. Every student of chemistry today knows that if you divide a drop of water into smaller and smaller droplets, you finally arrive at a single molecule of water, which, when further divided, is no longer water, but atoms of hydrogen and oxygen. This reawakening to atomic theory led to the rich field of Fundamental Particle Physics, which is one of the exciting frontiers of modern physics, and is described in Chaps. 8 and 9.

Quantum Mechanics carries the process of discretisation much further. In fact any electric charge must be a multiple of the charge of an electron (with the exception of quarks, whose charge, as we saw in Chap. 9, is a multiple of one third the electronic charge). The development of Quantum Mechanics in the 1920s was based on the realisation that not just matter, but also energy, and other quantities such as angular momenta, come in discrete packets, or quanta. Indeed, Einstein showed that matter and energy were different forms of the same thing, and could be transformed back and forwards from one to the other.

Around this time much debate was also taking place in mathematics on the nature of the continuum of real numbers. During the period 1918–1921, Hermann Weyl, a renowned mathematician and physicist, was tackling the problem of providing the mathematical continuum—the real number line—with a logically sound formulation. He came to accept that this aim was in principle impossible. As physics relies on mathematics for its formulation, one may well ask what implications such an observation might have for physics. In his moving tribute to his friend, Hermann Weyl, John Wheeler puts the question: *Do we not have to say that the notion of a physical world with a continuous infinity of degrees of freedom is an equal idealisation, an equal folly, an equal trespass beyond strict logic?* [20] In other words, shouldn't time and space also be quantised? The response to this question, like many others at the advancing edge of physics, depends on who is providing the answer [29].

In earlier Chapters, we have discussed several times the disagreement between the deterministic nature of Classical Physics and Relativity on the one hand, and the probabilistic nature of Quantum Mechanics on the other. If ever we are to develop a quantum theory of gravity, this conflict must be tackled. In a recent paper [30], Flavio Del Santo and Nicolas Gisin have cast doubt on the determinism of classical physics, claiming that *its alleged deterministic character is based on the metaphysical, unwarranted assumption of "infinite precision"*. By "infinite precision" they mean the specification of the value of a quantity as a real number (see Chap. 4) with an infinite number of decimal places.

As long ago as 1955, Max Born had raised similar doubts: "*Statements like 'a quantity X has a completely definite value' (expressed by a real number and represented by a point in the mathematical continuum) seem to me to have no physical meaning* [31]." In other words, a proposition, such as "X = 1", is neither true nor false, since X may equally well be equal to 0.999999999…, with any arbitrarily large number of digits "9".

Let us return to our question: are time and space also discrete? Apparently so, because for intervals of time below a certain threshold, it is difficult to conceive how any physical interaction can take place. The same considerations apply to space, since time and space must be treated on an equal footing, being part of the same four-dimensional space-time, as we saw in Chaps. 6 and 7.

As Del Santo and Gisin point out, whatever the size of these extremely small cells of space-time, they must be capable of carrying all the information about local interactions. In other words, information, and all the numerical quantities involved, must be embodied into a physical system (encoding), allowing it to be manipulated (computation) and transmitted (communication). But since most naturally occurring numbers have an infinite number of digits (even rational numbers such as 1/3 must be stored with an infinite string of digits), they cannot be stored in a finite (and extremely small) cell. They must somehow and somewhere be truncated. This implies that the mathematics becomes "quantized", since truncated numbers are multiples of some extremely small number, just as any charge is a multiple of the electronic charge.

Of course the above argument is by necessity oversimplified. Before proceeding one should "explain" quantitatively how the process of storing information works and how the stored information can be retrieved in order to influence the physics of the system. But for our purposes here it is sufficient to conclude that the truncation of numbers with an infinite number of digits can in principle destroy the determinism of most physical theories, and thus eliminate the main cause of incompatibility between QM and relativity.

Einstein's theory of General relativity, like the classical laws of Newton, is symmetric with respect to time. Both theories work equally well with time advancing from the present to the future, or retreating from the present to the past. One cannot tell by looking at a video of two balls colliding on a billiard table whether the movie has been run forwards or backwards. A few months before he himself died, Einstein wrote a letter to the family of a recently deceased, dear friend, Michele Besso, concerning his death: *"That signifies nothing. For us believing physicists, the distinction between past, present and future is only a stubbornly persistent illusion."* This symmetry between past and future does not exist in Quantum Mechanics where, as we have seen in the last Section, the act of observation plays such an important role in defining the present, and separating the past from what lies in the future.

So in these few paragraphs we have a brief glimpse into the fundamental contradictions between the two iconic theories of Twentieth Century physics. These issues are still very much under discussion, and as yet far from resolution. The search for a quantum theory of gravity, perhaps one that requires the discretisation of space and time, looks likely to continue apace for a considerable time yet.

12.10 Conclusion

In much ancient Greek and Roman theatre, the underlying structure of the play is basically the same: (a) introduce the situation; (b) make it as intricate as possible, and (c) find some unexpected and timely trick, capable of solving all the problems neatly, and tying up the loose ends. In a comedy, the good guys live happily ever after, and in a tragedy they usually die.

Sometimes it was the playwright's plot that got tied in knots, so that the only way he could extricate the protagonist from impending disaster was to bring a god on stage suspended from a crane. The deity promptly dispensed justice: the villains were punished and the good rewarded. Such a theatrical device became known as "*deus ex machina*", or "god out of a machine". Today it is a term applied to any highly contrived theatrical ending that does not arise from earlier events in the play.

Not all the ancients were enamoured of *deus ex machina*. Aristotle believed that: *it is obvious that the solutions of plots too should come about as a result of the plot itself, and not from a contrivance* [32]. Many centuries later, Anton Chekhov maintained that: *if in the first act you have hung a pistol on the wall, then in the following act it should be fired. Otherwise don't put it there* [33].

So what do these theories of dramatic structure have to do with our journey in this book? If all the world really is a stage, as Shakespeare stated [34], is there a storyline that science, and in particular, physics is following?

Up until Einstein's "magical" year of 1905, physics had proceeded more or less linearly, albeit with a few revolutions, such as Newton's unification of terrestrial and astronomical gravity, and Maxwell's unification of electricity and magnetism. Then along came Einstein and QM, followed by a "zoo" of unexpected elementary particles, and billions of galaxies, each containing billions of stars, to replace our all-inclusive solar system. To further increase the confusion, all these seemingly disparate fields (Relativity, QM, Particle Physics and Cosmology, plus others, such as Complexity Theory) seem to be for certain aspects intimately related, and for others, conflicting. The "cures" for this chaos (multiverses, String Theory, etc.), seem at times to be worse than the problems they are trying to solve, and we are flooded on a daily basis with new data, new conjectures, and new theories.

Our stage has become littered with various versions of Chekhov's gun, and we are not yet able to decide which ones will go off sometime in the future to clarify our storyline, and which ones, as was the case with the Steady State Theory of Cosmology, we may safely discard as being of no relevance to the plot.

In this book, and especially in this chapter, we have attempted to present a sample case of what is going on in physics, with no pretence that it is complete, nor that we have selected the most promising pathways, nor even presented the chosen ones in the right way. Other writers would have made their own, different choices.

Ultimately science is driven by the quest to determine the essence of our world: what is its origin? How does it work? What is its future? These mysteries are precursors to the ultimate question: what is the origin, purpose and ultimate fate of humanity, which has fascinated *homo sapiens* for thousands of years. Such questions have been tackled since the time of Pericles, Socrates, Plato and Aristotle using the tools of religion, philosophy and science.

In this book, we have adopted a scientific approach, since it is the only one which seems to us to have the potential of being truly universal, in contrast with the multiplicity of faiths and metaphysics, despite the continual disputes amongst scientists themselves. Actually, this bickering among peers can be regarded as beneficial, or even vital, to the progress of knowledge. However, we must acknowledge that we are still very far from any meaningful conclusions, and the plethora of Chekhov guns spread around our stage is confronting.

A subsidiary but still extremely important goal remains the quest for a Theory of Everything (TOE), capable of including all physical laws without any inconsistencies. Two opposite outcomes are possible:

1. No such theory will ever be found. As we have seen in this book, at least three times in the history of science, following Aristotle, Maxwell and Gellman (with the Standard Model of Fundamental Particles) it was thought the goal had been reached, or at least that we were close to it, and each time a multiplicity of exciting new physics was discovered. Perhaps we should learn from history, rather than being led astray by delusional hopes, and reconcile ourselves that a world, whose secrets we have been able to completely decipher, might seem "smaller" to many of us, and far less interesting.

2. With all the work going on in so many excellent research centres around the globe, perhaps we shall one day finally discover a *deus ex machina* to cast all of Chekhov's guns into oblivion, and reveal the answers to our questions with a compelling clarity. With Quantum Computing, and its incredible power, already coming into being, maybe we are nearer to this goal than we think.

Unfortunately, at the present time, observation and experiment are impractical in Fundamental Particle Physics at the energies necessary to test many theories, and in Cosmology for the earliest moments of the universe. This leaves us to rely on more tenuous and subjective qualities, such as beauty, to decide between competing ideas. This subjectivity leads to disagreement. We appear to be close, if not at, the limits of physics. To go further we risk passing into metaphysics, or religion.

With these words it is time to end our discussion. We began this book with the story of a robot, RT118/17, searching for an answer to the ultimate question of life, the universe and everything. It is therefore appropriate to conclude our book with the results of his journey, and the realisation that our own quest for knowledge will continue as long as curiosity remains a fundamental characteristic of the human condition.

12.11   RT and the Quantum Chip

Oil Can Harry, the Tier 1 robot, who ran the Lubrication Station near Sandsanrock Beach, had just completed his final grease job for the day, and was about to carry out the routine maintenance that kept his equipment in peak working order. His was the prime lubrication station in this part of the Dominion, and he was proud of its AAA rating. It was Harry's attention to detail that was largely responsible for its success.

Being only a Tier 1 bot had its advantages. His software did not allow him to become bored with routine tasks. His mind drifted back to the incident last year when a Tier 3 comrade had decided to walk into the nearby ocean to test out some theory about the meaning of life. Questions of this nature had never troubled him. He knew what his purpose in life was—it was printed on his duty statement—so he just got on with it.

Harry paused in his effort to disassemble the grease gun. Something was blocking out the light, and he turned to find the cause. To his surprise, there stood the very Tier 3 bot that had been occupying his thoughts. He recognised him only by the insignia RT118/17 etched into the carapace over where his heart would have been, if he'd had a heart.

Arty, as Harry had dubbed him, was still an imposing sight. However the arrogance, that last year had been so intimidating, was now gone, replaced by a remoteness, as though he now found his surrounds unworthy of his attention. This was reflected in his personal appearance. His carapace was coated in dust, intermingled with specks of what appeared to be rust. In short, he looked just plain shabby, and when he moved, his joints emitted a grating noise that sent Harry's auditory sensors into oscillation.

However, it was in his optics where the change was most noticeable: they no longer had the fierce stare that could bore through to your innermost chip. Instead, his gaze flitted over his surrounds with all the purpose of a drunken butterfly, as though disconnected from a mind that was totally engaged with deeper pursuits.

"Arty?" queried Harry. "It is you, isn't it?" The optics alighted for a moment on the lubricator's face, before resuming their spasmodic twitching search pattern. "What on earth has happened to you?" Harry recalled their conversation from last year. "Did you ever find the answer to that Ultimate Question?"

As though suddenly remembering where he was, RT focussed on the diminutive lubricator, and something of its former intensity returned to his stare. "Time. There's never enough time."

For one moment Harry thought his customer was about to storm out through the door in a mad dash to resume his quest. He sprang across to block his path, and ushered the creaking Tier 3 bot onto the lubrication bench, where he commenced the procedure for a full grease and oil job. "So what have you discovered then? About the meaning of life?"

"Life! Is a virus alive? Are we alive? What is life?"

"Well, of course we're – "

RT interrupted, his voice rising in excitement. "See that cupboard over there," he said, indicating a large cabinet for the storage of tools. "If you get inside there and shut the door, you become alive and dead."

"Alive or dead? Well, that's clear enough, isn't …"

"No! Alive *and* dead. At the same time!"

"This is all in the Master Data Repository?"

"Yes, and more. Much more." RT raised himself on an elbow to give the grease gun easier access. "Harry, you wouldn't believe …"

The lurching action revealed a tangle of fungus growing in the robot's armpit. It emitted an odour like the smell of decaying seagrass. "Well, haven't you let yourself go!"

RT began to tremble noticeably. It was the first time Harry had seen anything like it. Some part of the big bot's software was clearly malfunctioning. It must be his Emotions Simulation Chip.

"There might be more, you know. Ours may not be the only one."

"Only one what?"

"Universe. There could be trillions of others, and we're just one universe in some sort of bubble. Like a soap bubble. So many. So much more to be done." He started to get up. Harry pushed him back down.

"Lie still, Arty. I'm not finished yet."

RT tried to relax. It was an effort, but apart from an occasional twitch of his long legs he managed to succeed. "I've put in for an upgrade."

"What?" Harry couldn't imagine anyone taking such a step. He was perfectly happy the way he was, and certainly wouldn't want to go through another acclimatisation phase. The last one had been traumatic enough.

"There's a new technology coming out. Quantum based, using the principles of entanglement ..." He noticed Harry's blank expression. "Never mind. But it's fast. So many possibilities." His tremor returned, stronger than before, and he suddenly sat bolt upright, knocking Harry's grease gun to the floor. "I want it. I must have it." Harry picked up the gun and hung it on the tool rack. "I've sent in a request to the Dominion Robot Maintenance Centre."

"Wow! That's a big step."

"They can't refuse me. After all, I'm the Chief Operations Officer for the Master Data Repository."

"We're finished here, Arty. You're all set to go."

RT turned an impassioned face towards the low-tier bot, his ESC clearly in overdrive. "Don't you see, Harry? Until now, all I've done is read about what humans have achieved. With a quantum chip, it'll be like having a million brains all working together. I'll be smart. Really smart. I'll be able to work out new things for myself. So many new things."

"Like the meaning of life?

"Why not?" Harry gathered up the rest of his tools, and prepared to close the station for the night. "Well, maybe not that. But it doesn't matter. I can still be part of the quest. That's the important thing." He swung his feet to the floor, and headed for the door. "I told you before, Harry. It's my destiny. It's what I was made for."

He strode out through the door, and along the esplanade, with the late afternoon sun glinting off his newly shined carapace.

References

1. Alexander S (2016) The Jazz of physics. Basic Books
2. Shakespeare W, King lear, Act I, Scene I
3. Johnson S (2001) Emergence: the connected lives of ants, brains, cities and software. Allen Lane, The Penguin Press

4. Hossenfelder S (2018) Lost in math: how beauty leads physics astray. Basic Books
5. Liboff RL (1984) The correspondence principle revisited. Phys Today 37(2):50. https://doi.org/10.1063/1.2916084. Accessed 21 June 2020
6. Romero-Isart O, Juan ML, Quidant R, Cirac JI (2010) Toward quantum superposition of living organisms. New J Phys 12:033015. https://arxiv.org/pdf/0909.1469.pdf. Accessed 21 June 2020
7. Moreau PA et al (2019) Imaging Bell-type nonlocal behavior. Sci Adv
8. Schrödinger E (1944) What is life?
9. https://www.physicscentral.com/explore/action/pia-entanglement.cfm. Accessed 13 Aug 2020
10. Biello D. https://www.scientificamerican.com/article/when-it-comes-to-photosynthesis-plants-perform-quantum-computation/. Accessed 13 Aug 2020
11. Gilmore R (1995) Alice in Quantumland: an allegory of quantum physics. Copernicus Books
12. Wigner E (1960) The unreasonable effectiveness of mathematics in the natural sciences. Commun Pure Appl Math 13(I)
13. Hamming RW (1980) The unreasonable effectiveness of mathematics. Am Math Monthly 87(2):81–90
14. Breen PG, Foley CN, Boekholt T, Zwart SP (2019) Newton vs the machine: solving the chaotic three-body problem using deep neural networks (preprint, 2019). https://arxiv.org/pdf/1910.07291.pdf. Accessed 2 June 2020
15. Abbott D (2013) The reasonable ineffectiveness of mathematics. Proc IEEE 101(10)
16. Birkhoff G, Von Neumann J (1936) The logic of quantum mechanics. Ann Math Second Series, 37(4):823–843
17. Putnam H [1968] (1975) The logic of quantum mechanics. In: Putnam H (ed) Philosophical papers, vol 1, Mathematics, matter and method. Cambridge University Press, Cambridge, pp 174–97
18. Putnam H (2006) A philosopher looks at quantum mechanics (Again). Brit J Phil Sci 56:615–634
19. Putnam H (1994) Michael redhead on quantum logic. In: Clark P, Hale B (eds) Reading Putnam. Blackwell, Oxford, pp 265–80
20. Tegmark M (2003) Parallel universes in science and ultimate reality: from quantum to cosmos, honoring John Wheeler's 90th birthday. In: Barrow JD, Davies PCW, Harper CL (eds) Cambridge University Press, Cambridge; also Scientific American May, 2003. https://arxiv.org/abs/astro-ph/0302131. Accessed 22 June 2020
21. Davies P (2006) The Goldilocks Enigma: why is the universe just right for life? Allen Lane, London
22. Carr B, Ellis G (2008) Universe or multiverse? Astron Geophys 49(2):2.29–2.33. https://onlinelibrary.wiley.com/doi/pdf/https://doi.org/10.1111/j.1468-4004.2008.49229.x. Accessed 22 June 2020
23. Faulkner W, Requiem for a Nun

24. Wheeler JA (1986) Hermann Weyl and the unity of knowledge. Am Scientist 74(4):366–375. https://www.jstor.org/stable/27854250. Accessed 25 June 2020

25. The Matrix (1999) The Matrix Reloaded (2003) and The Matrix Revolutions (2003) A fourth film of the series is currently under production

26. Vedovato F et al (2017) Extending wheeler's delayed-choice experiment to space. Sci Adv 3(10):e1701180. https://advances.sciencemag.org/content/3/10/e1701180. Accessed 25 June 2020

27. Manning AG, Khakimov RI, Dall RG, Truscott AG (2015) Wheeler's delayed-choice Gedanken experiment with a single atom. Nat Phys 11:539–542. https://www.nature.com/articles/nphys3343. Accessed 25 June 2020

28. Fein YY et al (2019) Quantum superposition of molecules beyond 25 kDa. Nat Phys 15:1242–245. https://doi.org/10.1038/s41567-019-0663-9. Accessed 25 June 2020

29. https://www.scientificamerican.com/article/is-time-quantized-in-othe/. Accessed 4 July 2020

30. https://journals.aps.org/pra/abstract/https://doi.org/10.1103/PhysRevA.100.062107. Accessed 4 July 2020

31. Born M (1969) Physics in my generation. Springer, New York

32. Aristotle Poetics (1454a33–1454b9)

33. Gurlyand I (1904) Reminiscences of A. P. Chekhov. Teatr i iskusstvo 28(11):521

34. Shakespeare W, As You Like It, Act 2, Scene 7

Appendices

A.1.1 Zeno's Paradox

It is very easy to reject the conclusion of this paradox, but not so easy to refute it formally. In fact, the treatment of infinities was a tough nut for both mathematicians and scientists for many centuries, up to when modern infinitesimal Calculus, due to both Isaac Newton (1643–1727) and Gottfried Leibniz (1646–1716), provided a simple and very elegant tool for its general solution. In the case of Zeno's paradox, however, we can offer a simple arithmetical argument to explain it. Let us assume that Achilles needs, say, 1 min to cover the 100 m. He can then run the next 10 m in 0.1 min and the following one meter in 0.01 min and so on. Hence the total time to reach the tortoise will be the sum of all those time intervals, i.e. $1 + 0.1 + 0.01 + 0.001 + \ldots = 1.1111\ldots$, which is a finite quantity (equal to the rational number 10/9), in spite of being the sum of an infinite number of terms.

A.1.2 Frequency/Wavelength Limits for Human Sight and Hearing

Frequency and wavelength are reciprocal quantities. The range of wavelengths of the electromagnetic spectrum visible to the human eye, called the visible spectrum, runs between 0.39 and 0.7 μ (microns) approximately, where a micron is one millionth of a metre. It is bordered to the right by the *infrared* and to the left by the *ultraviolet*. The audible frequency range for the normal human ear lies between 20 and 20,000 Hz (Hertz).

A.2.1 Left Versus Right Hemispheres of Brain

There is now accumulating evidence dispelling the idea that some people are predominantly left-brained and others right-brained, which was always a demarcation found more in the popular media than the scientific literature. See, for instance: Nielsen et al. [1].

A.2.2 Second Law of Thermodynamics

The second law of thermodynamics received some notoriety in the 1960s when British scientist and acclaimed novelist, C.P. Snow, disparaged the lack of basic scientific knowledge among those who had received a classical education. He opined that ignoring the second law of thermodynamics was an equivalent in ignorance to never having read Shakespeare. Two satirists, Michael Flanders and Donald Swann, resolving to remedy the gap in their education, studied the law and based a song about it called: *The first and Second Law*, in their stage show, and in a later recording: *At the Drop of Another Hat*.

A.3.1 The Axioms of Euclid's Geometry

There are five axioms that are generally recognised as the basis for ordinary (or Euclidean) geometry. Let us blow away the cobwebs that have accumulated in our heads since high school, and recap these axioms in Figs. A.1, A.2, A.3, A.4 and A.5.

Fig. A.1 Axiom 1: A straight line can be drawn between any two points

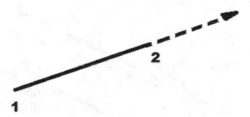

Fig. A.2 Axiom 2: Any terminated straight line can be projected indefinitely

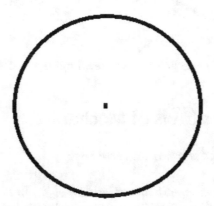

Fig. A.3 Axiom 3: A circle with any radius can be drawn around any point

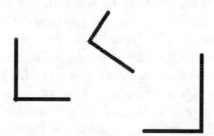

Fig. A.4 Axiom 4: All right angles are equal

From these humble beginnings the whole of Euclidean geometry can be derived by the strict application of logic.

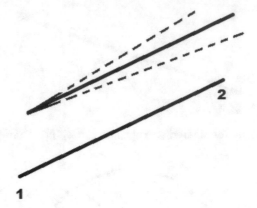

Fig. A.5 Axiom 5: Only one line can be drawn through a point parallel to another line.

A.3.2 Newton's Laws of Mechanics

Newtonian mechanics is based on three laws:

1. *An object at rest will remain at rest, and an object in motion will continue in uniform motion in a straight line, unless acted upon by an external force.* This law opposes the teachings of Aristotle who, as we saw in Chap. 2, believed that moving objects came to rest when the driving force was removed.
2. *The acceleration of an object produced by an external force is directly proportional to the magnitude of the force, in the same direction as the force, and inversely proportional to the mass of the object.* The astute reader will notice that in the case where the external force is zero, this law indicates that the acceleration of the object is also zero. The object therefore remains in its original state of rest or of uniform rectilinear motion. In other words, the first law is a special case of the second, and does not need to be stated separately.
3. *Action and reaction are equal and opposite.* (If you push on an object, it pushes back with an equal force.)

A.3.3 The Genius of Homer Simpson

Many of the script writers for the Simpsons are mathematicians, and some of their friends were physicists. There are two equations on the chalkboard referred to in Chap. 3. They are discussed by Simon Singh in his book: *The Simpsons and their Mathematical Secrets* [2].

The upper equation predicts the mass of the Higgs Boson (see Chap. 9) to fair accuracy *14 years before the particle was discovered*. This is quite an achievement for a cartoon character. The story is told in the link: [3].

The second relationship: $3987^{12} + 4365^{12} = 4472^{12}$ is, as noted in the text, difficult to disprove by evaluating the powers explicitly. However, considering only the last two digits of each number is sufficient for this purpose, as can be seen from the following:

A number abcdefg, where a, b, c, etc. are the digits forming the number, can be split into two parts: i.e., abcdefg = abcde00 + fg, where 0 is the digit zero.

Squaring the number gives: $abcde00^2 + 2.abcde00 \times fg + fg^2$.

Because of the zeros, the first 2 terms contribute nothing to the last two digits of the square, and so can be disregarded.

The same is true if we multiply two different numbers together (or add them): i.e. only the product (or sum) of the last two digits of each contributes to the last two digits of the product (or sum). Using this principle it is easy to show that the last two digits of the left hand side of the above relationship are not equal to the last two digits of the right hand side, and so the relationship is invalid.

However, if one were to actually evaluate the terms in the relationship fully, one would discover that $3987^{12} + 4365^{12}$ is close to the twelfth power of 4472.0000000070576171875, which a physicist would argue is *near enough* to 4472^{12}.

A.3.4 Principia Mathematica

Principia Mathematica is a very intense book by Alfred North Whitehead and Bertrand Russell, written in 3 volumes from 1910 – 1913. Their aim was to provide a formal logical derivation for all arithmetic. The uncompromising rigour of their approach is illustrated in Fig. A.6. This excerpt shows us that at this point in their opus (page 362), Whitehead and Russell have almost, but not quite, managed to prove that $1 + 1 = 2$.

Whitehead and Russell's book is the starting point for the work of Kurt Gödel, which we discussed in Chap. 4.

*54·43. $\vdash :. \, \alpha, \beta \, \epsilon \, 1 . \supset : \alpha \cap \beta = \Lambda . \equiv . \alpha \cup \beta \, \epsilon \, 2$

Dem.

$\vdash . \, \text{*54·26} . \supset \vdash :. \, \alpha = \iota^{\prime} x . \, \beta = \iota^{\prime} y . \supset : \alpha \cup \beta \, \epsilon \, 2 . \equiv . x \neq y .$

$[\text{*51·231}] \hspace{5cm} \equiv . \, \iota^{\prime} x \cap \iota^{\prime} y = \Lambda .$

$[\text{*13·12}] \hspace{5cm} \equiv . \, \alpha \cap \beta = \Lambda \hspace{1cm} (1)$

$\vdash . (1) . \, \text{*11·11·35} . \supset$

$\hspace{1.5cm} \vdash :. \, (\exists x, y) . \, \alpha = \iota^{\prime} x . \, \beta = \iota^{\prime} y . \supset : \alpha \cup \beta \, \epsilon \, 2 . \equiv . \alpha \cap \beta = \Lambda \hspace{1cm} (2)$

$\vdash . (2) . \, \text{*11·54} . \, \text{*52·1} . \supset \vdash . \, \text{Prop}$

From this proposition it will follow, when arithmetical addition has been defined, that $1 + 1 = 2$.

Fig. A.6 Extract from Whitehead and Russell's *Principia Mathematica*

A.3.5 Falsifiability

The concept of falsifiability, or the notion that every physical theory must be capable, at least in principle, of being proven wrong, was introduced by the philosopher of science, Karl Popper. In Part 3, we see that there are currently diverging opinions on whether a lack of falsifiability is sufficient grounds for rejecting a physical theory. Also in Chap. 4, we see that the logician, Kurt Gödel, showed that certain propositions inside a system cannot be proved from within the system itself. In addition, undecidable statements, such as the famous paradox *This sentence is false*, obviously cannot be falsified.

A.3.6 Variation of the Fine Structure Constant

A very tight bound on current variations of the Fine Structure Constant α has been set at one part in 10^{17} per year [4]. This result however does not preclude past variations. Research by a group at the University of New South Wales has shown a variation in α of approximately one part in 10^{5} spanning $\sim$23% to 87% of the age of the universe [5]. More recent work by this group claims that the changes in α are different in different directions in the universe [6].

A.4.1 Russell's Paradox

Russell discovered the paradox that bears his name in 1901 while using set theory to formalise the mathematics of arithmetic. A *set* can be considered to be a collection, e.g. the set of all dogs is a collection that contains all dogs and nothing else.

Russell explored the question of whether a set can contain itself. In the above example, this is clearly not the case; a set of dogs (i.e., a collection of dogs) is not a dog. However, if we consider a set of sets, i.e., a collection of collections, then it is possible for the set to contain itself. In fact, the set of *all* sets *must* contain itself.

Suppose now we take this process one step further, and consider *the set of all sets that do not contain themselves.* The paradox occurs when we ask ourselves: "Does this set, so defined, contain itself, or not?" If we say, "no, it does not contain itself", we see from its definition as the set of all sets that do not contain themselves, that it *does* contain itself. Conversely, if we say that it does contain itself, we see that this is wrong, because by definition it is the set of all sets that do not contain themselves, and therefore must not contain itself.

If you are having trouble getting your head around the paradox, spare a thought for the innocent eight-year-olds of the sixties, who were expected to master set theory on their way to learning that $2 + 2 = 4$.

A.4.2 Gödel's Incompleteness Theorem

In analogy with the Liar's Paradox that we have already discussed, let us consider the statement: *This statement is unprovable.* If the statement is provable, then the proof will prove something that is false, which does not bode well for mathematics. The only alternative is that the statement is unprovable. In other words, although the statement is true, it cannot be proved, i.e., there are some statements that, although true, cannot be proved.

Of course, the above argument is only an outline of the logic that Gödel employs in his proof. His genius lies in encoding *This statement is unprovable* into a natural number, and then examining the implications within the rules of arithmetic. For a deeper insight into this intriguing topic, please see *Gödel's Proof* by Ernest Nagel and James R. Newman, referenced in Chap. 4. A summary of Gödel's approach is given by Natalie Wolchover [7].

A.5.1 Common Sense

The remark: *common sense is not so common* is often attributed to Voltaire: *Le sens commun n'est pas si commun* in the *Dictionnaire philosophique portatif*,

1764. However, much earlier, a Roman poet, Decimus Iunius Iuvenalis (*aka* Juvenal) in Book III of his *Satires* had written *Rarus enim ferme sensus communis – Common sense is generally rare."* Much later, Mark Twain and Will Rogers added their own contributions: *I've found that common sense ain't so common.* In any case, common sense, based as it is on our past experiences in a macroscopic world, turns out to be a poor guide in QM.

A.5.2 Black Body Radiation

A *black body*, when cold, is a perfect absorber of radiation, i.e. all radiation that falls on it is absorbed; none is reflected or re-emitted. When heated, a black body is the most efficient emitter of radiation allowed by the laws of physics. Normal objects, such as iron bars or lamp filaments, are less efficient radiators than black bodies.

A black body can be approximated in the laboratory by hollowing out a cavity in a metal ingot. Light falling into the cavity is repeatedly reflected around inside it before being finally absorbed. (A similar effect happens with narrow-necked bottles that are used in summer to trap flying insects.) As none of the light that enters escapes, the cavity appears black. However, if the ingot is heated, the cavity appears to be brighter than the metal around it, because the cavity is a better emitter than the surrounding metal.

A.5.3 Nature Does not Make Jumps

Natura not facit saltus or *Nature does not make jumps* was a truism of classical science, attributed by some to Gottfried Leibniz, co-founder of the calculus. It is the belief that nature does not allow sudden changes. In biology, it was used by Charles Darwin to support his theory of the evolution of species through small gradual changes, rather than through the sudden appearance of new species. In physics, Quantum Mechanics, with its sudden transitions between states, violates this principle. Also in biology, the discovery of the roles of DNA and mutations in genetics shows that sudden changes, even if small, are indeed possible.

A.5.4 Laplace's Demon

The following observation by French polymath Pierre-Simon, Marquis de Laplace (1749 – 1827), introduced what is now known as Laplace's demon: *We may regard the present state of the universe as the effect of its past and the cause of its future. An intellect which at a certain moment would know all forces that set nature in motion, and all positions of all items of which nature is composed, if this intellect were also vast enough to submit these data to analysis, it would embrace in a single formula the movements of the greatest bodies of the universe and those of the tiniest atom; for such an intellect nothing would be uncertain and the future just like the past would be present before its eyes* [8].

Even before the advent of QM, Laplace's demon had been disputed by physicists, using thermodynamical arguments. However, these refutations have themselves been criticised, and debate still continues.

A.5.5 The Monty Hall Problem

When this problem was presented in a popular magazine in the U.S., many mathematicians wrote rude letters to the editor complaining about the published solution, and how it demonstrated the deplorable lack of numeracy prevailing in the U.S. Unfortunately for them, they were wrong and the magazine solution was right.

Statisticians tackling the problem would probably use Baye's Theorem for conditional likelihoods. However, the rest of us can use the following argument. When the contestant makes a choice for the door the car is behind, she has a 1 in 3 chance of being right, and a 2 in 3 chance of being wrong. No matter what the host does, this does not change. There is always a 2 in 3 chance that the car does not lie behind the door that the contestant has chosen. When the host, who knows where the car is, opens one of the other doors and there is no car there, the contestant knows that there is now a 2 in 3 chance that the car is behind the other unopened door. She should therefore change her selection.

If the reader still has difficulty believing this result, a simple experiment should suffice to convince them. Take three cups and a coin. Turn your back, and ask a friend to conceal a coin beneath one of them. Guess which cup hides the coin, then ask your friend to turn up one of the other two cups that does not hide the coin. Check whether your initial choice was right, or whether you would have been better off to change your selection. Repeat the experiment many times, and keep a tally of the results. You will find that you are twice as likely to win if you change your mind.

A.6.1 Origin of Relativity Limerick

A letter by A. H. Reginald Buller to the Journal of the Royal Astronomical Society of Canada in 1938 explains the origin of this much-cited limerick.

"As the author of the relativity limerick, perhaps I may be allowed to say that the limerick was made by me about fifteen years ago while sitting in the garden of my friend and former colleague, Dr. G. A. Shakespear, Lecturer on Physics at the University of Birmingham.

After conversing together on Einstein's theory, I suggested that we should each try to make a relativity limerick. At the end of about five minutes, the limericks were ready and were exchanged, but with nothing more than a trace of mutual admiration."

A.6.2 A Land of Bliss Poem

A Thousand Reflections Quiver in These Lofty Boughs;
The Flowers of the Grotto Greet the Arriving Guest.
Near the Spring, Where then is the Herb Gatherer?
A Lone Boatman Rows on the Stream,
And His Guitar Sounds Two Notes.
The Boat Glides Lazily, the Gourd Offers Its Wine.
Shall We Ask the Boatman of Vo Lang:
"Where Are the Peach Trees of the Land of Bliss?
by George F. Schultz [9]

A.7.1 Precession of Perihelion of Mercury

The rotation of the major axis of the orbit of mercury is known, in astronomical parlance, as *the precession of the perihelion of mercury*, where *perihelion* is the term for the point of closest approach of the orbit to the sun.

The observed rate of this precession can be measured very precisely, and is found to be 5599.7 arcseconds of rotation per century, again not an effect that can be observed in the backyard. Almost all (5030 arcseconds/century) of this figure can be accounted for by Newton's theory, and another 530 arcseconds/century is caused by perturbations to the orbit from the outer planets. This leaves a 40 arcseconds/century discrepancy, which is very well explained by General Relativity, when one includes the distortion of space–time caused by the gravitational mass of the sun.

A.7.2 Gravitational Redshift of Light

What happens when light tries to escape from a really intense gravitational field, such as that in the neighbourhood of a black hole? In Chap. 7, we discussed the event horizon that surrounds a black hole, and saw that nothing, not even light, can exit from within this region. Another way of explaining this is to say that the gravitational redshift of light is so severe that the light is redshifted to zero frequency. Zero frequency light, of course, does not exist.

As mentioned in Chap. 7, the first reliable verification of the gravitational redshift was made in 1954 by Daniel M. Popper [10]. He utilised the line spectra (see Chap. 8) emitted by a hot gas, and measured the frequency of 42 hydrogen lines on 27 spectrograms of the white dwarf star, 40 Eridani B. He obtained a frequency shift of 0.007 percent, which was within experimental error of the value predicted by General Relativity.

So sensitive have modern experiments become that it is no longer necessary to use astronomical observations to test General Relativity. Instead, atomic clocks have been flown around the world, and rockets launched tens of kilometres into space to study the time dilation caused by gravity. Such an experiment in 1976 found General Relativity to be accurate to within 0.007% [11].

A.7.3 Gravitational Lensing

Suppose that we have a situation where a distant star or galaxy lies behind a region where the gravitational field is intense. Light passing through this field will be deflected, and as a consequence the distant object will appear to lie in a different location from its true position.

This effect, known as *gravitational lensing*, is illustrated in Fig. A.7. Light from the source S passing above the massive object G will appear to be coming from location A, while light passing below G will appear to originate at location B. Fig. A.7 is a two-dimensional representation of a scenario

Fig. A.7 Light from source S is bent by the gravitational field at G, so that to an observer located at O, the light appears to come from somewhere on the circle, or Einstein Ring, joining A and B

that in reality is three dimensional. Light could also have passed in front of, or behind, the massive object. The net result is that light from the faraway source appears to originate from anywhere on the circumference of a circle around the source, depending on which path the light happened to take.

A.7.4 Evidence of Gravitational Waves from Binary Pulsar Systems

When stars that are 4 to 8 times the size of our sun reach the end of their lives after having consumed all of their nuclear fuel, they undergo spectacular death throes. With the pressure exerted by outflowing radiation no longer sufficient to oppose internal gravitational forces, such stars collapse under their own weight. Energy released by the collapse results in the outer layer of the star being blown away in an enormous supernova explosion. The core collapses further until all atomic structure disappears, and protons and electrons combine into neutrons,[1] leaving behind what is now known as a neutron star. A pulsar is a rotating neutron star.

As we saw in Chap. 7, neutron stars may be observed as isolated objects, or in some cases, binary systems, where they rotate about another object, in which case astronomers are able to calculate their mass. As such a system loses energy, the two components of the binary system rotate more closely about each other, i.e., the radius of their orbit decreases, as they fall slowly together. Hence, the time to complete each orbit, i.e. the *period of rotation*, decreases as the objects move faster and faster on their paths around each other. This is the same effect that causes an ice skater to spin faster when she pulls her arms close to her body during a spin, and is a consequence of one of the most far-ranging laws in physics: the law of conservation of angular momentum.

In 1974, the discovery of the Hulse-Taylor pulsar, which is located 21,000 light years from earth in a binary system with another neutron star, enabled a

---

[1]These sub-atomic particles are discussed in Chap. 9.

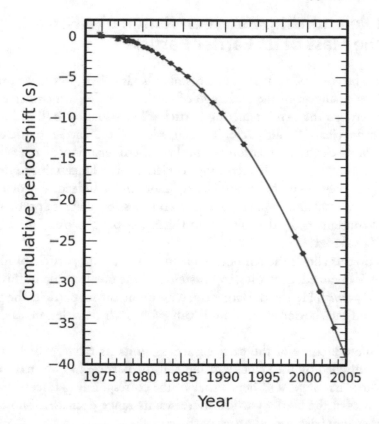

Fig. A.8 The decay of the orbital period of the binary system containing the Hulse-Taylor pulsar. The continuous line is the expected trend predicted from General Relativity. The diamonds are the measured values. Image: public domain (Image: Inductiveload—Own work, Public Domain https://commons.wikimedia.org/w/index.php?curid=9538634 (accessed 2020/7/28))

further test of General Relativity. The period of rotation of the binary system has been measured accurately over several decades. The results are presented in Fig. A.8.

As can be seen from the figure, the period of rotation of the binary system has decayed, precisely according to the predictions of General Relativity, over thirty years of measurement. The confidence of physicists in the reality of gravitational waves therefore grew, and motivated the construction of even more sensitive apparatus in an attempt to detect these elusive ripples in space–time on their passage through the earth. As discussed in Chap. 7, the first direct detection of a gravitational wave was made on 14th September, 2015, and with improvements in the technology since that date, detections have become almost commonplace.

A.9.1 Relationship Between Range of a Force and the Mass of Its Carrier Particle

The force between two particles at a distance is visualised, in Quantum Field Theory, as arising from the exchange of carrier particles, which carry *information* between the two interacting particles. The situation is illustrated, for example, in Chap. 9, Fig. 9.5c. The emission of the carrier particle from the proton is clearly a violation of the law of conservation of mass/energy. Such violations are possible according to Heisenberg's Uncertainty Principle, but only if they occur for a sufficiently short time. Indeed, a violation of mass/energy by an amount ΔE *for a time* Δt is possible if Δt is approximately Planck's constant h divided by 4π and ΔE. *During this time, the violation cannot be detected.*

This process allows the temporary creation of a carrier particle of mass m, where m is obtained, according to Einstein and the one formula in this book, from $\Delta E = mc^2$. Hence, the larger the mass of the carrier particle, the greater is ΔE, and the shorter is the time it can exist. Such particles are said to be *virtual*.

The finite lifetime of the carrier particle limits its range, which is given by Δt multiplied by the velocity of light. If the particle has no mass, e.g., it is a photon, its range, and hence that of the corresponding force is infinite. As the mass of the carrier particle increases, its range decreases. Short range forces therefore have heavy carrier particles.

A.9.2 Table of Quarks

In the table below, which at first glance bears a strong resemblance to our earlier table for leptons, we summarise what is known about quarks. It should be noted that the masses of the quarks are notoriously hard to pin down [12].

Gen	Name	Symbol	J	B	Charge[a]	Iso-spin	C	S	T	B′	Anti-quark	Mass[b]
1	Up	u	1/2	1/3	2/3	1/2	0	0	0	0	$\bar{u}$	0.0021
1	Down	d	1/2	1/3	-1/3	-1/2	0	0	0	0	$\bar{d}$	0.0051
2	Charm	c	1/2	1/3	2/3	0	1	0	0	0	$\bar{c}$	1.36
2	Strange	s	1/2	1/3	-1/3	0	0	-1	0	0	$\bar{s}$	0.098
3	Top	t	1/2	1/3	2/3	0	0	0	1	0	$\bar{t}$	185
3	Bottom	b	1/2	1/3	-1/3	0	0	0	0	-1	$\bar{b}$	4.45

[a]Charges are expressed relative to the magnitude of the electronic charge
[b]Masses are expressed relative to the proton mass

Here "Gen" is the Generation, J is the angular momentum (spin) of the quark, B the baryon number, C the charm, S the strangeness, T the topness and B' the bottomness. The charge is the electrical charge in units of one electron charge. It can be seen, as was mentioned earlier, that the charges of the quarks are 1/3 and 2/3 that of the electronic charge. The charges, and also the baryon numbers with their values of 1/3, give a clue that baryons are most likely constructed from three quarks. This is indeed true. The antiquarks have the opposite quantum numbers to the quarks, and are given the names: antiup, antidown, etc. Mesons are formed from a quark-antiquark pair, and are consequently bosons.

A.9.3 Grand Unified Theory (GUT) and Supersymmetry

Grand Unified Theory (GUT)

In this theory, the three interactions we have discussed thus far (electromagnetic, weak and strong) are merged at high energy into a single force. At lower energies, the theory should result in the same predictions as the current Standard Model. However, at high energies, new particles are predicted. Unfortunately, the energies required to produce these particles are well above the capabilities of modern accelerators, and as a consequence they have not been observed.

In the absence of this type of direct evidence, support for the GUT is sought from indirect observations of physical quantities, such as proton decay and the electric dipole moments of particles, where the predictions of the Standard Model are inaccurate. Some versions of the GUT predict the existence of the magnetic monopole.

All observed magnetic fields arise from pairs of north and south magnetic poles. These are known as magnetic dipoles. A magnetic monopole would comprise an isolated north (or south) pole, in the same way that electric charges (positive or negative) can exist in isolation. Due to the lack of experimental confirmation of its predictions, there is currently no generally accepted GUT.

Supersymmetry

The distinction between fermions (particles of half odd-integer spin) and bosons (carrier particles of zero or integer spin) may seem rather arbitrary. What if each fermion had a *supersymmetric* partner with identical quantum numbers, but with integer spin? These supersymmetric particles (called spar-ticles) would comprise sleptons, squarks, neuralinos and charginos with spins differing by ½ from the normal fermions. Unfortunately, once again these new particles are so heavy that they are beyond the energy range of today's particle colliders.

A.10.1 The Measurement of Distance in Astronomy

The measurement of distances in astronomy is fraught with difficulties. Nevertheless, it is common to see in newspapers, popular science magazines, and books such as this one, that objects have now been discovered at a distance of 10 billion light years. How are such measurements even possible?

We have already explained in Chap. 10 how nearby stars observed from earth appear to move against the background of distant stars as the earth circles the sun, and how the observed parallax leads to the measurement of distance. The standard unit of distance in astronomy is the parsec; a star with one arc-second of observed parallax, measured when the earth is on opposite sides of the sun, is said to be at a distance of 1 parsec. (An arc-second is 1/3600 degrees.) The distance to any star in parsecs is the reciprocal of its measured parallax in arc-seconds. To convert from parsecs to units more familiar to non-astronomers (kilometres or light-years), we need to know the distance of the earth from the sun. This distance is defined as the Astronomical Unit (AU) and must be measured independently.

It is generally recognised that the first successful measurement of the Astronomical Unit was carried out in 1672 by Jean Richer and Giovanni Domenico Cassini, who measured the parallax to the planet Mars from two different locations on earth. Knowing the separation of these terrestrial loca-tions provides a baseline length for the parallax measurement, and enables the distance from earth to Mars to be calculated. Kepler's Laws of planetary motion give the interplanetary distances in terms of the Astronomical Unit, so working backwards, a knowledge of the earth-Mars separation can be used to estimate the value of the Astronomical Unit. The currently accepted value is: 149,597,870,691 ± 30 m. From this value for the AU, 1 parsec is found to be about 3.3 light-years.

In the early '90s, the Hipparcos satellite was used to take parallax measure-ments of nearby stars to an accuracy much greater than possible with earth-bound telescopes, thereby extending the distance measurements of these stars out to ∼ 100 parsecs (∼300 light-years). How can this range be further extended?

The first step involves the comparison of a star's known luminosity with its observed brightness. The latter is the brightness (or apparent magnitude) observed at the telescope. It is less than the intrinsic luminosity (or absolute magnitude) of the star because of the inverse-square attenuation with distance of the observed radiative energy. If we know how bright the star is intrinsically, we can estimate its distance using the inverse square law.

From measurements on stars in our galaxy within the 300 light-year range it was discovered that stars of similar particular types have the same absolute magnitude. These stars are dubbed *standard candles*. If stars of these types are observed outside the range where parallax measurements are observable, we can estimate their distance by assuming their absolute magnitude is the same as closer stars of the same type, observe their apparent magnitude, and use the inverse square law to compute their distances. The distance to another galaxy can be inferred from the distance to particular stars within it.

The term "standard candle" may sound strange in this context. A standard candle was originally defined as a one-sixth-pound candle of spermaceti wax, burning at the rate of 120 grains per hour, and was used when comparing the intensity of light sources. Its output was one candlepower, a unit of luminous intensity now obsolete, much to the relief of the oceans' sperm whales. In astronomy, the term "standard candle" is used to designate a class of objects whose members have a fixed intrinsic brightness. Two examples are Cepheid Variable stars and supernovas of a particular type (Type Ia).

Supernova explosions are rare events. The last observation of a type Ia supernova explosion in our galaxy was made by Johannes Kepler, and others, on the 9th October, 1604. Earlier, in 1054 AD, Chinese astronomers had observed another famous supernova explosion, the remnants of which now make up the Crab Nebula. The extreme brightness of supernovas—whose peak light output can equal that of the entire galaxy that contains them—enables us to observe them in distant galaxies, and thereby estimate the distance to these galaxies.

A.11.1 Measurement of the Curvature of the Universe

The temperature fluctuations visible in the Cosmic Microwave Background, and described in Sect. 11.5, can be understood by considering the plasma that existed before the Recombination Era. The ingredients in this "soup" were electrons, atomic nuclei, photons, dark matter and various baryons, but no stable atoms, as the high temperature would have stripped the electrons from any atom that chanced to form. Occasional density fluctuations would occur, both higher and lower than the average density. The regions of high-density dark matter would tend to attract particles due to increased gravitational attraction. As plasma flowed into these regions it would become compressed.

Compressed plasma has a high internal pressure due to the electromagnetic interactions. Once the pressure had increased, it would drive the particles apart, lowering the plasma density in that region, thereby enabling the dark matter, by means of its gravitational attraction, to pull more plasma into the region, and begin the cycle over again. What we have therefore are propagating periodic density fluctuations, which are very similar in concept to sound waves in air. As a consequence physicists call these plasma oscillations "baryon acoustic oscillations".

The speed of these oscillations (sound waves) through the plasma is estimated to be 60% of light speed. As the age of the universe at the time of the Recombination Era was 380,000 years, the maximum distance that any oscillation could have travelled in this time is 0.6 × 380,000 light years (LY), i.e. ~ 230,000 LY. This is the largest distance over which oscillations could interfere with each other. For distances separated by more than this distance, the universe had been in existence for insufficient time for oscillations to have passed between the separated points. As a consequence this upper limit of 230,000 LY is known as the "sound horizon".

As we saw in Sect. 11.4, a frequency analysis of the Cosmic Microwave Background (CMB) reveals the fluctuations present at the time of the Recombination Era. The oscillation with the largest spatial extent is the one corresponding to the sound horizon. However, it is not the absolute size of the sound horizon that can be measured in this way, but only its angular size, i.e. the angle it subtends in the sky. Assuming that space–time is flat, and thus that the light rays are not bent but linear, a fluctuation in the CMB the size of the sound horizon would subtend an angle of one degree when observed from earth. If space–time is not flat but curved, the angle subtended would be more or less than one degree, depending on the nature of the curvature. The measurements of the CMB obtained from the Planck observatory show that the universe is topologically flat at large scales to within 0.5%. This unexpected result is explained in Sect. 11.5.

References

1. Nielsen JA, Zielinski BA, Ferguson MA, Lainhart JE, Anderson JS (2013) An evaluation of the left-brain versus right-brain hypothesis with resting state functional connectivity magnetic resonance imaging. Published: August 14, 2013, https://doi.org/10.1371/journal.pone.0071275. Accessed 7 July 2020
2. Singh S (2013) The simpsons and their mathematical secrets. Bloomsbury
3. https://boingboing.net/2014/10/17/homers-last-theorem.html. Accessed 9 July 2020
4. Rosenband T et al (2008) Frequency ratio of Al+ and Hg+ single-Ion optical clocks. In: Metrology at the 17th decimal place science 319(5871):1808–1812
5. Webb JK, Murphy MT, Flambaum VV, Dzuba VA, Barrow JD, Churchill CW, Prochaska JX, Wolfe AM (2001) Further evidence for cosmological evolution of the fine structure constant. Phys Rev Lett 87:091301
6. Webb JK., King JA, Murphy MT, Flambaum VV, Carswell RF, Bainbridge MB (2011) Indications of a spatial variation of the fine structure constant. Phys Rev Lett 107:191101
7. https://www.quantamagazine.org/how-godels-incompleteness-theorems-work-20200714/#jump2. Accessed 22 July 2020
8. Laplace P-S (1951) A philosophical essay on probabilities, English Trans. by Truscott FW, Emory FL, Dover Publications, New York, p 4
9. Schultz GF (1968) Vietnamese legends. Charles E. Tuttle Publishing Company
10. Popper DM (1954) Red shift in the spectrum of 40 Eridani B. Astrophys J 120:316
11. Vessot RFC et al (1980) Test of relativistic gravitation with a space-borne hydrogen maser. Phys Rev Lett 45(26):2081–2084
12. https://phys.org/news/2010-05-masses-common-quarks-revealed.html. Accessed 30 July 2020

Index

Printed in the United States
By Bookmasters